WHAT A

Marcus Chown is an awar merly a radio astronome ogy in Pasadena, he is weekly science magazin selling *We Need to Tall* *neory Cannot Hurt You*, *The Never-Ending Days of Being Dead* and *The Magic Furnace*. He also wrote the children's book *Felicity Frobisher and the Three-Headed Aldebaran Dust Devil* and *The Solar System*, the bestselling app for iPad, which won the Future Book Award 2011.

Further praise for the books of Marcus Chown:

'Marcus Chown has a talent for explaining complex ideas through simple analogies, and his enthusiasm is infectious.' Tim Radford, *Guardian*

'We must all be grateful to writers like Chown who are able to make accessible work that in its crude form is not only inaccessible to outsiders, but unknown to them.' *Independent on Sunday*

'[Chown] superbly catches the spirit of scientific adventure.' *Sunday Times*

'Marcus Chown rocks!' Brian May

'Chown writes with ease about some of the most brain-bending concepts.' BBC *Focus Magazine*

'Staggeringly good.' Rufus Hound

'He makes you hug your knees, and rock back and forth saying "Whoa!"' *Dazed and Confused*

'Chown writes very fluently, helping us visualise things with matchboxes and Lego bricks.' Steven Poole, *Guardian*

WHAT A WONDERFUL WORLD

*Life, the Universe and Everything
in a Nutshell*

MARCUS CHOWN

FABER & FABER

First published in 2013
by Faber and Faber Limited
Bloomsbury House
74–77 Great Russell Street
London WC1B 3DA
This paperback edition published in 2014

Typeset by Agnesi Text
Printed in the UK by CPI Group (UK) Ltd, Croydon, CR0 4YY

A CIP record for this book
is available from the British Library

ISBN 978-0-571-27841-1

2 4 6 8 10 9 7 5 3 1

To Jeanette, Karen & Aline. So glad I found you.
With love, Marcus

CONTENTS

Foreword xi

PART ONE: HOW WE WORK

1 : I am a galaxy 3
   CELLS

2 : The rocket-fuelled baby 21
   RESPIRATION

3 : Walking backwards to the future 33
   EVOLUTION

4 : The big bang of sex 49
   SEX

5 : Matter with curiosity 69
   THE BRAIN

6 : The billion per cent advantage 87
   HUMAN EVOLUTION

PART TWO: PUTTING MATTER TO WORK

7 : A long history of genetic engineering 105
   CIVILISATION

8 : Thank goodness opposites attract 121
   ELECTRICITY

9 : Programmable matter 139
   COMPUTERS

10 : The invention of time travel 151
   MONEY

11 : The great transformation 165
   CAPITALISM

PART THREE: EARTH WORKS

12 : No vestige of a beginning 181
   GEOLOGY

13 : Earth's aura 193
   THE ATMOSPHERE

PART FOUR: DEEP WORKINGS

14 : We are all steam engines 213
   THERMODYNAMICS

15 : Magic without magic 225
   QUANTUM THEORY

16 : The discovery of slowness 247
   SPECIAL RELATIVITY

17 : The sound of gravity 259
   GENERAL RELATIVITY

18 : The roar of things extremely small 279
   ATOMS

19 : No time like the present 299
   TIME

20 : Rules of the game 311
   THE LAWS OF PHYSICS

PART FIVE: THE COSMIC CONNECTION

21 : The day without a yesterday 323
   COSMOLOGY

22 : Masters of the Universe 341
   BLACK HOLES

Acknowledgements 353
Permissions 354
Notes 355
Bibliography 397
Further Reading 403
Index 405

FOREWORD

This book came about because I have an exceptional editor – Neil Belton. In fact, I am his stalker. I have pursued him all the way from Jonathan Cape to Faber. Neil has many talents. But one of those talents is that he knows what his authors are good at and what they should be writing better than they do.

My skill is that I can take complex physics and explain it to someone on a number 25 bus (or perhaps I should say someone *unfortunate enough* to be sitting next to me on a number 25 bus). But, in addition to physics, I am also interested in other things. I read a lot of fiction. I am interested in history. I like running. In fact, in 2012, I did the London Marathon (something I rarely fail to mention in the first three minutes of meeting someone).

Neil's big idea was that I combine these two things: that I use my skill at explaining complex physics in layperson's terms to explain *everything* in layperson's terms.

I was daunted. How could I possibly write about everything? Where would I even start? I began thinking about how to organise such a wide range of material logically. But I tore up outline after outline. What changed everything, however, was writing *Solar System for iPad*. I had only 9 weeks to write 120 stories about planets, moons, asteroids and comets, so I had no option but simply to dive in and learn to swim on the job. It must have worked because the App won several awards. So that is what I did. I overcame my apprehension and just dived in.

It was a struggle. Usually, when I need to know something about physics, I identify a physicist – it could be a Nobel Prize-winner – and simply phone them. There is a 95 per cent chance they will be able to answer my stupid questions immediately. And, if they cannot, they will at least make an attempt at answering them. But, with subjects I knew nothing about, such as money, sex and the human brain, it was difficult even to identify someone who might be able to answer my incredibly basic questions. And, when I did and phoned them, they were often not able to explain things at the toddler level I needed. Worse, it was sometimes as if we were speaking different languages. Often, I had to go to two, three or four people before I could find someone who could answer all my questions. And, on occasion, I could not find anyone who was able to do that. Instead, I was forced to piece together an explanation from things people I had gone to had said and from things I had read.

But Neil was right. This was the book I should have been writing. It was one that stretched me beyond my comfort zone and that, ultimately, proved to be an exhilarating and a joyful experience. I loved learning about all kinds of things I know nothing about. And I began to appreciate what a wonderful world we live in – one far more incredible than anything we could possibly have invented. Along the way, I learnt many surprising things, such as . . .

- To understand a single collaterised debt obligation squared – one of the toxic investments that sunk the world economy in 2008 – would require reading *1 billion pages* of documentation
- Slime moulds have 13 sexes (and you think you have problems finding and keeping a partner)

- You could fit the whole human race in the volume of a sugar cube
- You are ⅓ mushroom – that is, you share ⅓ of your DNA with fungi
- You age more slowly on the ground floor of a building than on the top floor
- The crucial advantage that humans had over Neanderthals was . . . *sewing*
- IBM once predicted that the global market for computers was . . . *five*
- Today your body will build about 300 billion cells – more than there are stars in our Galaxy (no wonder I get knackered doing nothing)
- Believe it or not, the Universe may be a giant hologram. *You* may be a hologram

If everything in our information-overloaded society has passed you by in a high-speed blur, my book just might bring you quickly and painlessly up to speed on how the world of the twenty-first century works. It is, after all, one man's attempt to understand everything. No, I cannot really claim that. It's one man's attempt to understand everything . . . *volume one*.

Marcus Chown, London, March 2013

PART ONE: How we work

I:

I AM A GALAXY

Cells

A good case can be made for our non-existence as entities.

LEWIS THOMAS

There's someone in my head and it's not me.

PINK FLOYD

I think I am me. But I am not. I am a galaxy. In fact, I am a thousand galaxies. There are more cells in my body than there are stars in a thousand Milky Ways. And, of all those myriad cells, not a single one knows who I am or cares. *I* am not even writing this. The thought was actually a bunch of brain cells – neurons – sending electrical signals down my spinal cord to another bunch of cells in the muscles of my hand.[1]

Everything I do is the result of the coordinated action of untold trillions upon trillions of cells. 'I like to think my cells work in *my* interest, that each breath they draw for *me*, but perhaps it is *they* who walk through a park in the early morning, sensing my senses, listening to my music, thinking my thoughts,' wrote American biologist Lewis Thomas.[2]

The first step on the road to discovering that each and every one of us is a super-colony of cells was the discovery of the cell itself. Credit for this goes to Dutch linen merchant Antonie van Leeuwenhoek. Aided by a tiny magnifying glass he had adapted from one used to check the fibre density of fabrics, he became the first person in history to *see* a living cell. In a letter published in April 1673 in the *Philosophical Transactions* of the Royal Society of London, van Leeuwenhoek wrote, 'I have observ'd taking some blood out of my hand that it consists of small round globuls.'

The term 'cell' had actually been coined two decades earlier by the English scientist Robert Hooke. In 1655, he had examined plant tissue and noticed dead compartments stacked together. However, neither he nor van Leeuwenhoek realised that cells are the Lego bricks of life. But that is what they are. A cell is the 'biological atom'. There is no life – as far as we know – *except cellular life.*

Prokaryotes: a protected micro-universe

The first evidence of cells comes from fossils about 3.5 billion years old. But there is more tentative evidence, from about 3.8 billion years ago, in the shape of telltale chemical imbalances in rocks that are characteristic of living things. The first cells, known as prokaryotes, were essentially just tiny transparent bags of gloop less than a thousandth of a millimetre across. The bag, by concentrating stuff inside, speeded up key chemical reactions such as those that generate energy. It also protected proteins and other fragile products of those reactions from toxic substances such as acids and salt in the environment. The bag of gloop was an island haven in an ocean of disorder and chaos, a protected micro-universe where order and complexity might safely grow.

The complexity of such cells was in large part due to the proteins – megamolecules assembled from amino-acid building blocks and made of millions of atoms. Depending on their shape and chemical properties, these Swiss-army-knife molecules can carry out a myriad tasks, from speeding up chemical reactions to acting as cellular scaffolding to flexing like coiled springs to power the movement of cells. Even a simple bacterium possesses about four thousand different proteins, though some proteins, such as those needed for reproduction, are assembled, or expressed, only intermittently. The structure of these proteins is encoded by deoxyribonucleic acid, or DNA, a double-helical molecule floating freely as a loop in the chemical soup, or cytoplasm, inside a cell.

Cellular structure is beautifully intricate. First, there is the bag, or membrane. This is made of fatty acids, molecules that are characterised by having a water-loving end and a water-hating end. When such lipids come together in large numbers – typically a billion – they spontaneously self-assemble into two layers, with their water-hating ends on the interior and their water-loving ends on the outside.

The lipid layers that enclose a cell are not a passive barrier. Far from it. This double skin regulates the molecules coming in and going out of the cell. Imagine the cell as an ancient city surrounded by a wall. In the same way that small creatures such as mice can pass easily back and forth through the city wall, small molecules can pass unhindered in and out of the cell membrane. And, just as bigger creatures such as people are admitted only through gates in a city wall, the passage of big molecules is regulated by 'gates' in the cell membrane. For instance, there are proteins shaped like hollow tubes spanning the width of the membrane through which bigger molecules can tunnel into and tunnel out of the cell. And there are transporter proteins whose job is to shuttle bigger molecules physically from one side of the membrane to the other.

The molecules that come in to the cell are those needed for energy and to make proteins and to provide information about the outside world. For instance, an abundance in the surrounding environment of molecules necessary for building new cells might trigger a cell to reproduce.[3] On the other hand, a shortage of water molecules coming across the membrane might warn a cell that it is in danger of drying out. This might trigger a cascade of chemical reactions inside the cell, ultimately causing a stretch of DNA to be copied repeatedly into molecules called ribonucleic acid, or RNA. These find their way to ribosomes, nanomachines that use the RNA templates to make proteins that might be components of a mucus that will protect the cell from dehydration.[4] Too big to pass through the cell membrane, the proteins flooding out through the cytoplasm in their millions are packaged into membrane sacs, or vesicles, which fuse with the cell membrane. The membrane can then open up, without rupturing and losing its structural integrity, and cast them into the outside world.

But cells, in addition responding to molecules in their environment, also respond to molecules from *other cells*. Even the simplest and most ancient prokaryotes cooperated with each other, which is revealed by fossils of large microbial communities known as stromatolites. Living stromatolites can still be found today – for instance, in shallow tropical waters off the western coast of Australia – but the oldest of these fossil communities is about 3.5 billion years old.

At the same time that a cell makes proteins to protect itself from environmental changes, it might produce proteins that warn others of its kind to do the same. Such chemical signalling is crucial to the survival of simple prokaryotes, which often live in huge colonies known as biofilms, quite possibly the first organised structures to appear on Earth. The cells on the inside of such a biofilm might secrete a sugary protein that sticks their membranes to the membrane of other cells, whereas those on the outside of the film might produce proteins that help protect them from environmental toxins. Some cells will even kill themselves in order to yield up precious nitrogen for the good of their companions. This kind of cooperation, with cells within a group differentiating to carry out different tasks, is reminiscent of the cells in our bodies. It hints at how such cellular super-cooperation might have got started billions of years ago.

There are limits on the size and complexity of prokaryotes. For one thing, the proteins assembled, or expressed, by their DNA can travel only by drifting slowly, or diffusing, across a cell. Beyond a certain size, a prokaryote is therefore suicidally slow in reacting to environmental dangers. This problem has been solved by rare prokaryotes such as *Thiomargarita namibiensis*, discovered only in 1997. The giant sulphur bacterium, which is about 0.75 millimetres across and easily visible to the naked eye, possesses not one loop of DNA but *thousands*, spread evenly throughout its cytoplasm. This means that proteins expressed by local strands of DNA, even if they diffuse slowly, can still get to all parts of the cell rapidly.

But there is another serious problem that keeps prokaryotes small. The bigger one of them grows the more energy it needs. If it were to use the strategy of *T. namibiensis*, however, an increasing proportion of that energy would be needed for manipulating large quantities of DNA. Since this would be at the expense of any other cellular processes, the road to increased complexity is well and truly blocked.

But there is another way to grow big: take up cannibalism.

Eukaryotes: cities in bags

About 1.8 billion years ago, a prokaryote swallowed another prokaryote. Prokaryotes actually include bacteria and more exotic archaea bacteria, microorganisms that survive in extreme environments such as boiling sulphur springs and so were probably among the first life forms on Earth.[5] What actually happened 1.8 billion years ago was that an *archaeobacterium swallowed a bacterium*.

Such an event must have occurred innumerable times before. But, in all cases, the bacterium was either devoured or spat out. This time, for some unknown reason, the bacterium survived. More than that. *It thrived*. There was some mutual benefit for the swallower and swallowee. The latter found a protective environment, safe from the hostile outside world, and the former a new power source.

The evidence that something like this did indeed happen was gathered by the American biologist Lynn Margulis (the first wife of TV astronomer Carl Sagan). And the evidence is still around us today. The energy-generating mitochondria inside the eukaryotic cells of all animals are not only the same size as free-living bacteria but they *look like them too*.[6] Even more striking, they have their own DNA, which is separate and distinct from the DNA of the whole cell, and fashioned into a loop exactly as in prokaryotes.

In fact eukaryotes may have hundreds, or even thousands, of such mitochondria. These are self-contained power stations, furiously reacting hydrogen from food with oxygen to make life's mobile power packs, adenosine triphosphate, or ATP.[7] 'My mitochondria comprise a very large proportion of me,' wrote American biologist Lewis Thomas. 'I suppose there is almost as much of them in sheer dry weight as there is of the rest of me. Looked at this way, I could be taken for a very large, motile colony of respiring bacteria.'[8]

With a cell's mitochondria working semi-autonomously in this way, it is no longer necessary for it to devote so much of its DNA to the task of generating energy. The DNA is free to encode other things, other protein nanomachinery. Consequently, when cells gained mitochondria 1.8 billion years ago, they were suddenly free to grow a whole lot bigger and more complex.

A large eukaryote compared with a typical prokaryote is like a cat beside a flea. Such a mega-cell might contain hundreds, even thousands, of membrane-enshrouded bags. These organelles divvy up the chores of the cell, functioning as the equivalent of factories, post-office sorting offices and other specialist buildings in a modern-day city.

Lysosomes, for instance, are the garbage-disposal units of the cell. They break down molecules such as proteins into their building blocks so they can be used again. The reason the lettuce in your burger wilts is that heat from the beef breaks down the lysosome membranes of the lettuce cells. This unleashes enzymes, which devour the lettuce. Other organelles include the rough endoplasmic reticulum, which acts like a cellular DHL office. Dotted with ribosomes, it translates RNA arriving from the nucleus into proteins destined for foreign parts beyond the cell. Yet another organelle is the Golgi apparatus, which acts like a packaging centre. It can modify proteins, wrapping them, for instance, in a sugar coating that absorbs water. Such proteins can be used to make the surfaces of blood cells slimy so they can move about more easily.[9]

In fact, a eukaryotic cell is less like a single organism than a colony of organisms, each of which long ago lost its ability to survive alone. 'For the first half of geological time our ancestors were bacteria,' says Richard Dawkins. 'Most creatures still are bacteria, and each one of our trillions of cells is a colony of bacteria.' And all of this has come about by chance. 'The mitochondrion that

first entered another cell was not thinking about the future benefits of cooperation and integration,' says Stephen Jay Gould. 'It
was merely trying to make its own living in a tough Darwinian
world.'[10]

The organelles are subservient to the cell's nucleus, which contains its DNA and orchestrates pretty much all cellular activity.
The English botanist Robert Brown recognised the nucleus as a
common feature of complex cells in 1833.[11] Enclosed in a double
membrane, the nucleus is reminiscent of a walled castle inside
the walled city of the cell. The membrane controls the passage
of molecules into the nucleus and the passage of proteins expressed by the DNA out of the nucleus.

The presence of a nucleus is one of the defining features of a
eukaryote, along with the presence of a plethora of organelles.
A prokaryote has neither a nucleus nor organelles. In fact, the
very word *prokaryote* means 'before kernel, or nucleus', while
eukaryote means 'true nucleus'. Very probably, a nucleus is a
necessity in a cell as complex as a eukaryote because of the need
to protect the precious DNA from the frenzied activity going on
in every corner.[12]

In addition to having a nucleus and a large number of
organelles, a eukaryote contrasts with a prokaryote in having a
cytoskeleton. Proteins such as tubulin form long scaffolding poles
that criss-cross the cell. Such microtubules stiffen the soft bag of
the cell, giving it a shape. They also anchor organelles to the
membrane. This ensures that they are arranged in a similar way
in all eukaryotes much as internal organs are arranged in a similar
way in all humans. But, in addition to providing internal scaffolding, microtubules act as an internal rail network that can
rapidly transport material about the cell. They do this by growing

at one end and disintegrating at the other, so, bizarrely, *it is the track rather than the train that provides the motive power.* Newly made proteins, enclosed in bags, or vesicles, simply hop on a convenient microtubule and are instantly speeded off to a far-away destination within the cell.

The cellular rail network enables a eukaryote to overcome one of the biggest obstacles preventing a prokaryote becoming big: getting stuff around the cell. A eukaryote, rather than having to wait for proteins to diffuse slowly through the cytoplasm, speeds them around on its rapid transit network.

But eukaryotes, despite being an enormous advance over prokaryotes, also have their limits. Orchestrating organelles is a complex activity. If a cell contained more than a few thousand of them, such orchestration would be beyond the capability of a nucleus. Eukaryotes, like prokaryotes, are a biological dead end. The way to increasing complexity lies in another direction – in cooperation on an unprecedented scale.

Multicellular organisms

From the moment they arose, eukaryotes almost certainly cooperated with each other in increasingly sophisticated ways. But, about 800 million years ago, they crossed a critical threshold. Nature had put together colonies of symbiotic prokaryotes to make eukaryotes. Now it repeated the trick. It put together colonies of symbiotic eukaryotes to make multicellular organisms.

The fact that life on Earth spent about 3 billion years at the single-cell stage before it took the step to the multicellular stage is probably telling us that the step is a difficult one. This has implications for the prospects of finding extraterrestrial life.

Despite fifty years of searching, astronomers have seen no sign of intelligence elsewhere in our Galaxy. One possibility is that life is common in the Milky Way but only in the form of single-celled microorganisms.

Humans – as well as animals, plants and fungi – are all multi-cellular organisms. Each of us is a colony of about 100 million million cells. They come in about 230 different types, ranging from brain cells and blood cells to muscle cells and sex cells, and all are enclosed in a bag made of skin cells, no less a container than the membrane of a single cell.

Each cell has its own copy of the same DNA (apart from blood cells in their mature form, which are so utilitarian they lack even a nucleus). But whether a cell becomes a kidney cell or a pancreatic cell or a skin cell depends on the particular section of the DNA that is read, or expressed. This, in turn, depends on regulatory genes – themselves stretches of DNA – which can turn off and turn on the reading of DNA, depending on things such as the concentration of a particular chemical in the locality.

Each of the 100 million million cells that makes up a human being is a micro-world as complex as a major city, buzzing with the ceaseless activity of billions of nanomachines. It has store-houses, workshops, administrative centres and streets heaving with traffic. 'Power plants generate the cell's energy,' says American journalist Peter Gwynne. 'Factories produce proteins, vital units of chemical commerce. Complex transportation systems guide specific chemicals from point to point within the cell and beyond. Sentries at the barricades control the export and import markets, and monitor the outside world for signs of danger. Disciplined biological armies stand ready to grapple with invaders. A centralised genetic government maintains order.'[13]

And all of this is going on every moment of every day of our lives while we remain utterly oblivious to it. In the words of biologist and writer Adam Rutherford, 'Each movement, every heartbeat, thought, and emotion you've ever had, every feeling of love or hatred, boredom, excitement, pain, frustration or joy, every time you've ever been drunk and then hungover, every bruise, sneeze, itch or snotty nose, every single thing you've ever heard, seen, smelt or tasted *is your cells communicating with each other and the rest of the Universe.*'[14]

We all start our lives as a single cell when a sperm, the smallest cell in the body, fuses with an ovum, the biggest cell in the body and one actually visible to the naked eye. Every human in fact spends about half an hour as a single cell before it splits into two. This is a phenomenal process in itself. In a mere thirty minutes, not only must a cell make a copy of its DNA – a process that, for speed, occurs simultaneously at multiple sites on the DNA – but it must construct something like 10 billion complex proteins. This is more than *100,000 a second*.

Within sixty minutes, the two cells split into four, then later eight, and so on. After several divisions, chemical differences across the developing embryo cause the cells to differentiate. It is a process that culminates in cells 'knowing' they have to be kidney cells or brain cells or skin cells. Over years, a single cell becomes a galaxy of cells – or, rather, a *thousand galaxies of cells*.

Hardly any of the cells in your body – apart from brain cells – are permanent. The cells lining the wall of the stomach are bathed in hydrochloric acid strong enough to dissolve a razor blade, so must be remade constantly. You get a new stomach lining every three or four days. Blood cells last longer but even they self-destruct after about four months. It is fair to say that you are pretty much a new person every seven years, something that maybe explains the seven-year itch. You look at your partner and suddenly think, 'That's not the person I got together with!'

The cells of your body die in such prodigious numbers that, simply to replace them, you must build about 300 billion new cells every day. That is more cells than there are stars in our Galaxy. No wonder it can be tiring doing nothing.

Aliens

There may be an astronomical number of cells in your body. But they are not able to carry out all the functions necessary for your survival. Not without assistance from legions of alien cells such as prokaryotes, fungi and single-celled animals called protozoans.[15] In your stomach, for instance, hundreds of species of bacteria work constantly to extract nutrients from your food. If some of these 'good' bacteria are inadvertently killed by antibiotics, the result can be an affliction such as diarrhoea.

The alien bacteria protect you from illness by filling niches in your body that otherwise might be filled by disease-causing pathogens. The Human Microbiome Project, a five-year study funded by the US government, presented its findings in 2012. It found that the nasal passages of about 29 per cent of people contain *Staphylococcus aureus* – better known as the MRSA superbug. Since such people suffer no ill effects, the implication is that in healthy people the bugs act as good bacteria, keeping harmful pathogens at bay.

Remarkably, the Human Microbiome Project found that there are more than 10,000 species of alien cells in your body – 40 times the number of cell types that actually belong to you. You are only 2.5 per cent human. In fact, about 5 million bacteria call every square centimetre of your skin home. The most densely populated regions are your ears, the back of the neck, the sides of the nose and your belly button. What all these alien bacteria are doing is a mystery. The Human Microbiome Project found that 77 per cent of the species in your nose, for instance, have completely unknown functions.

The sheer number of alien bacteria in your body might actually underrate their importance. The Human Microbiome Project found that microorganisms that inhabit your body have a total of at least 8 million genes, each of which codes for a protein with a specific purpose. By contrast, the human genome contains a mere 23,000 genes.[16] Consequently, there are about 400 times as many microbial genes exerting their effect on your body as human genes. In a sense, you are not even as much as 2.5 per cent human – you are merely 0.25 per cent human.

Since the alien cells in your body are largely prokaryotes, which are much smaller than eukaryotes, they add up to a few kilograms or a mere 1–3 per cent of your mass. They are not encoded by your DNA but infected you after birth, via your mother's milk or directly from the environment. They were pretty much all in place by the time you were three years old. It is fair to say that we are born 100 per cent human but die 97.5 per cent alien.

The biological event horizon

Every cell is born from another cell. '*Omnis cellula e cellula*', as François-Vincent Raspall first recognised in 1825. Consequently, every cell in our body – every cell on the Earth – can trace its ancestry in an unbroken line back to the very first cell, which appeared about 4 billion years ago. The first cell is generally referred to as the last universal common ancestor, or LUCA. Nobody knows how exactly it came about. Undoubtedly, there was a vast amount of experimentation – a huge amount of pre-evolution – before nature hit on the design.

Mistakes, or mutations, in genes accumulate at a steady rate over time. So, if one species has twice as many mutations of a particular gene as a second species, we can say it split from a common ancestor twice as far back. This is how the tree of life, first envisaged by Charles Darwin, is constructed. However, bacteria have an inconvenient habit of swapping DNA as well as passing DNA to their descendants. This means that, in the vicinity of LUCA, the tree of life is less a tree and more like an impenetrable thicket.

In physics, scientists talk of the 'event horizon' of a black hole – the point of no return for infalling matter. It cloaks the black hole so that nothing can be seen of its interior. Similarly, biologists talk of the biological event horizon beyond which nothing can be known. There, unfortunately, lies LUCA.

Since the time of LUCA, the Earth, despite dabbling in multi-cellularity, has been a bacterial world. There are believed to be something like 10,000 billion billion billion bacteria on our planet. That is a billion times more bacteria than there are stars in the observable Universe. But this might not give a true picture of terrestrial biology. Consider viruses. 'We live in a dancing matrix of viruses,' wrote Lewis Thomas. 'They dart, rather like bees, from organism to organism, from plant to insect to mammal to me and back again, and into the sea, tugging along pieces of this genome, strings of genes from that, translating grafts of DNA, passing around heredity as though at a great party.'[17] Incapable of reproducing without hijacking the machinery of cells, viruses are generally not considered to be precursors of cellular life. But who knows?

2:

THE ROCKET-FUELLED BABY

Respiration

All our energy is a beam of sunlight set free from
its captive state in food.
NICK LANE, *Life Ascending*

We make our living by catching electrons at the
moment of their excitement by solar photons,
swiping the energy released at the instant of
each jump and storing it up in intricate loops for
ourselves.
LEWIS THOMAS, *The Lives of a Cell*

A rocket rises on a column of white smoke and orange flame. A baby kicks out in a moment of joy. These two things may appear to have nothing in common. But appearances are deceptive. Both are energised by essentially the same chemical reaction. Both are powered by rocket fuel.

A moment's thought shows why this is not surprising. Boosting a heavy rocket into space requires the most powerful fuel – the one that, pound for pound, packs the biggest oomph. Life on Earth has been engaged for 3.8 billion years in trial-and-error experimentation. It would be odd if, in its efforts to power living organisms, it too had not stumbled on the most potent available energy source.

That energy source is the chemical reaction between hydrogen and oxygen. In the case of all animals, it is hydrogen extracted from food and oxygen extracted from the air. In the case of a rocket, it is liquid hydrogen and liquid oxygen.

So how does the reaction between hydrogen and oxygen work? And where exactly does the tremendous energy come from? That requires a little background knowledge.

All atoms, including those of hydrogen and oxygen, consist of a tiny nucleus and even tinier electrons. The electrons orbit the nucleus, snared by its powerful electric force in much the same way that planets, influenced by the force of gravity, orbit the Sun. There are many different ways the electrons can orbit in a given atom. But, in general, they are happiest being as close to the nucleus as possible to minimise their energy.

This is a general principle of physics. For instance, a ball high on a hillside is said to have high gravitational energy. Given the slightest opportunity, it will try to minimise its energy – that is, roll down to the bottom of the hill where it has low gravitational energy. Similarly, the electrons in an atom, as surely as balls rolling downhill, will try to minimise their energy.

When two atoms come together, there may be new ways for their combined electrons to arrange themselves. If there is a configuration with a lower total energy than in the two separate atoms, then, as inevitably as a ball rolling downhill, the atoms will combine to form a molecule. This is all chemistry is: the rearrangement of electrons.

Since the energy of the molecule is less than the energy of the separate atoms that came together to make it, there is energy left over. It is a cornerstone of physics that energy can be neither created nor destroyed, only transformed from one form to another – for instance, from electrical energy into light energy. Consequently, the surplus energy becomes available to *do* things.[1]

In a rocket, for instance, the reaction between a hydrogen atom and an oxygen atom – actually, *two* hydrogen atoms react with each oxygen atom to make H_2O (water) – liberates a large amount of energy. This heats the water and expels the white vapour at great speed from the back of the rocket. Action and reaction being equal and opposite, the high-speed exhaust propels the rocket forward.

The liberation of so much energy by the reaction between hydrogen and oxygen is the reason it can lift a rocket to the edge of space.[2] It is the reason why a marathon runner can go 26 miles

385 yards on a bowl of pasta. It is the reason why the reaction is exploited by every last animal on Earth.

Actually, the reaction between hydrogen and oxygen is not the only one that liberates energy. Before oxygen was present in substantial quantities in the Earth's atmosphere, organisms gleaned their energy from much less efficient processes such as fermentation. Yeast cells make alcohol via fermentation. The muscles of sprinters, when they run short of oxygen, make lactic acid via fermentation. In the fermentation process only about 1 per cent of the surplus energy is made available to do work. This can be compared with a whopping 40 per cent in the case of the reaction of hydrogen with oxygen.

These two numbers tell us something interesting and profound about the biological world. In order to have carnivores, it is necessary to have at least three layers in the food chain: plants, animals that eat plants, and animals that eat animals that eat plants. But, if only 1 per cent of the energy of plants is available to the animals that feed on them, then only 1 per cent of 1 per cent – that is a mere 0.01 per cent – is available to animals that feed on those animals, and so on.

So until oxygen became available in more-or-less modern quantities about 580 million years ago, there could be no carnivores. (Actually, bacteria learned the oxygen trick more than 2 billion years ago but there were only tiny amounts of O_2 available at the time.) In fact, it is estimated that the oxygen trick boosted the amount of biomass on Earth by an astonishing factor of about 1,000. Instead of two tiers, or trophic levels, levels in the food chain, suddenly it was possible to have five or six. The bewildering complexity of life on Earth today owes everything to the exploitation of oxygen.

Battery-powered biology

But how exactly does the oxygen trick work? In a rocket, hydrogen and oxygen combine to make water, with the explosive release of a large amount of heat energy. Clearly, living organisms do not make use of such a violent process. They would be blown apart. Instead, they liberate the energy, step by step, in a far less destructive and subtle way.

What actually happens when hydrogen and oxygen react together in a rocket is what happens in all chemical reactions: the electrons play a game of musical chairs. Specifically, an oxygen atom grabs electrons from two hydrogen atoms.[3] In the process, the oxygen and hydrogen atoms fuse into a molecule of water.[4] However, say the hydrogen atoms supply electrons to an oxygen atom *but the hydrogen atoms and oxygen atom never actually meet?* This is the non-explosive twist on the oxygen–hydrogen reaction that is exploited by biology.

The first requirement is to obtain hydrogen. The gas does not exist in its free state on Earth. Being the lightest of all gases, if created in any quantity, it would float off into space. Inside a cell, however, an amazingly subtle and energy-efficient process called the Krebs cycle strips hydrogen atoms from food – that is, from either molecules of sugar (glucose – $C_6H_{12}O_6$) or fat. Two hydrogen atoms then donate their electrons to an oxygen atom. Only this does not happen directly, as in a rocket. Between the hydrogen atoms and the oxygen atom stretches a long wire made of protein complexes.[5] And the donated electrons, bursting with excess energy, hop from location to location down the wire.

Focus on a single electron. As it hops down the wire, as inevitably as a ball rolling downhill, it drives hydrogen nuclei, or protons,[6] through channels, or pores, in the cell membrane.[7] Since protons carry an electric charge – the opposite of electrons – this charges up one side of the membrane with respect to the other. Something like this happens in a battery; it creates an electric force field between the battery's terminals. And, actually, this hints at what the super-energetic electron does as it thunders down the protein wire to an oxygen atom: it turns the cell membrane into a charged-up battery. The resulting electric force field across the membrane is stupendously powerful. It is comparable, in fact, to the electric field that, in a thunderstorm, breaks down the atoms in the air and unleashes a multimillion-volt bolt of lightning.[8]

You might imagine that the cells in your body should crackle with lightning. However, the tremendous electric force field extends over only the tiny thickness of a cell membrane – about 5 millionths of a millimetre – and other molecules intervene to stop this force field having its way. Interestingly, however, in programmed cell death, or apoptosis, this protective mechanism is turned off and cells are in effect killed off by their own internal lightning bolts.

The powerful electric force field of the membrane battery drives a chemical reaction that creates adenosine triphosphate, or ATP. Such molecules are stores of energy; think of them as portable batteries. So, as the electron bounces down the protein wire, losing energy all the while, it leaves in its wake a large number of energy-packed ATP molecules. Released into the wild, these have the ability to power cellular processes wherever and whenever necessary.

In the final analysis, you are battery powered. There are about a billion ATP molecules in your body, and all of these are used and recycled every 1–2 minutes. Toys may require a handful of batteries that become flat in a few hours. Contrast this with your body, which uses up 10 million power packs *every second*. Thank goodness that, for human beings, batteries are included.

Finally, the electron arrives at the end of the protein wire, exhausted of its energy. There it combines with the waiting oxygen atom. When a second electron from another hydrogen atom joins it, the oxygen atom achieves the highly desirable state of a filled outer shell of electrons. But this is not quite the end of the story. If the oxygen atom passes the electrons to a carbon atom – left behind when hydrogen was stripped from the food in the Krebs cycle – the result is a very stable molecule of carbon dioxide. And carbon dioxide, along with water vapour, is what oxygen-breathing animals exhale as waste.

Breathing

So much for the chemistry of respiration; what about its physiology? Well, we breathe in air, of which about 20 per cent is oxygen. Only about a quarter of this actually gets used, so exhaled air still contains about 15 per cent oxygen. This is why it is possible to revive an unconscious person with exhaled breath via mouth-to-mouth resuscitation.

Our breath is drawn deep down into our lungs, the inner surfaces of which have a structure on the smallest scale rather like a branching tree. The branches, known as alveoli, run along-side fine blood vessels, and oxygen molecules pass from them to red blood cells. The tree-like structure of the alveoli maximises the area over which this oxygen transfer can occur, maximising the amount of oxygen that can enter the blood stream. Remarkably, the surface area of a human lung is similar to that of a tennis court.

When an oxygen molecule is transferred to a blood cell, it is picked up by a giant protein called haemoglobin. It is then ferried to a cell where the oxygen is combined with hydrogen stripped from food to liberate energy for the cell. Crucially, haemoglobin changes its behaviour depending on the acidity of its surroundings. The acidity at its cellular destination changes the protein in a subtle way so that it *repels* rather than attracts its passenger. The protein therefore drops off its precious oxygen molecule. But the change in the haemoglobin means that it now *attracts* a molecule of carbon dioxide. As soon as one latches on, it is promptly ferried back to the lungs, where it passes from blood vessels to alveoli and is exhaled.

The oxygen we breathe and that powers all the biological processes in our bodies is essential to keep us alive. Whereas we can survive without food for a month, and without water for a week, we can survive with our air supply cut off for only about three minutes.[9] Every instant of our lives we are a mere three minutes from death. This fact becomes shockingly apparent to a heart-attack victim whose heart stutters to a halt and stops pumping oxygen around the arteries and blood vessels of his or her body.[10]

Photosynthesis

But where does the oxygen we breathe come from? The answer, of course, is plants. Rather than breathing in oxygen and breathing out carbon dioxide, they take in carbon dioxide and pump out oxygen.

Pretty much all the energy used by life on Earth is therefore ultimately the energy of sunlight, which plants capture directly from the Sun.[11] The trick is mind-bogglingly clever – otherwise we would long ago have found a way of mimicking it and powering human civilisation directly from sunlight. The energy of a particle of light – a photon – is transferred to an electron in a giant protein called chlorophyll. This is the molecule responsible for the green pigment of plants, although life also uses a second, non-green version. Bursting with energy, the electron can then energise chemical processes. Photosynthesis is a bewilderingly complex process but, in essence, it achieves the exact opposite of respiration.

Whereas respiration splits hydrogen from foods such as sugars and passes its electron to oxygen, ejecting as waste carbon dioxide, photosynthesis splits hydrogen from water and uses it with carbon from carbon dioxide to build sugars, ejecting as waste the leftover oxygen. Pause for a moment to think what an amazing trick this is. Using nothing more than water, carbon dioxide from the air and sunlight, plants are able to synthesise energy-rich food.

The sugars made by plants are in essence captured sunlight. And, whenever we eat plants, we in effect unleash the energy of this captured sunlight. But the miracle does not end there. Some plants such as trees, when they die, can become buried and transformed by heat and pressure deep down in the ground into fossil fuels such as coal. When we burn coal, we unleash yesterday's sunlight. Ultimately, everything on Earth is powered by a beam of captured sunlight.

Photosynthesis is actually quite inefficient. The percentage of incoming light energy that is converted into sugar in most plants is only about 1 per cent. The race is on, therefore, not only to create artificial photosynthesis but to make it *significantly better* than in nature – converting say 20 per cent of incident sunlight into hydrogen.

Hydrogen when combined with oxygen – think rocket fuel, think respiration – liberates large amounts of energy. It could therefore be used, in fuel cells, to power all manner of machines from cars to computers. There are three main steps in the creation of artificial photosynthesis. First, light must be captured and its energy transferred to an electron, boosting its energy. Next, the electron must be freed from its parent atom. Finally, the super-energetic electron must be used to smash apart water to liberate the all-important hydrogen. Artificial photosynthesis, able to make hydrogen fuel from sunlight, would wean the human race from its dependence on fast-dwindling reserves of fossil fuels such as oil. It would be a game-changing technology. It could transform the world.

3:

WALKING BACKWARDS TO THE FUTURE

Evolution

Evolution is a tinkerer.
FRANÇOIS JACOB, 'Evolution and Tinkering'

Pigs look us straight in the eye and see an equal.
WINSTON CHURCHILL

Question: What do aeroplanes and television sets and lamp posts have in common with frogs and whales and people? Answer: All are highly improbable configurations of matter and all do what they do extremely well. The technological things in the first group were designed by human beings. An obvious conclusion to draw from the similarity between the two groups would therefore be that the living things in the second group were also designed. The obvious conclusion, however, is wrong.

The illusion of design in nature is so strong that it was not recognised as an illusion until the nineteenth century. In Europe at the time, there was a pretty much universal belief that living things had been created and put on the Earth in their present forms by a Supreme Being. The scientists of the day were mostly religious and the very last thing they wanted to do was question such an idea and bring down on themselves the wrath of the Church. However, scientists have no choice but to go with the evidence. And the evidence was overwhelming that the bewildering diversity of life on Earth – everything from bacteria to blue whales, fungi to flying foxes, gorillas to giant sequoias – is the consequence of a purely natural mechanism.

An important clue came from fossils. These appeared to be the relics of ancient creatures, buried by sediments settling to the bottom of lakes and seas, and somehow – nobody knew exactly – turned to stone. Fossils reveal that the creatures that inhabit the

Earth today are not the same as the ones that once inhabited the planet. Some ancient creatures such as the dinosaurs have disappeared entirely whereas other vanished creatures appear related to creatures today. The simplest, most primitive creatures appear fossilised in the oldest sediments. As the rock layers became progressively younger, the fossils became ever more complex and sophisticated.

The idea dawned on scientists that the fossil record was a *time sequence* of life on Earth. It was telling us that, over vast tracts of time, species of creatures gradually change their appearance, morphing from one into another and eventually becoming the species we see around us today. Life was not created on Day One by a Creator, remaining frozen and static forever after. Instead, it has evolved, gradually, from simpler ancestral forms.

Such evolution explains the striking similarities between creatures living today such as humans and chimpanzees. If all life on Earth descended from a common ancestor in the distant past, it is obvious that all creatures today are related. But what drives evolution? What causes species to change over the generations? And how have all creatures ended up doing what they do so incredibly well that they give every appearance of being designed? The man who found the answer was Charles Darwin.

Darwin embarked on HMS *Beagle* in 1831. During his five years as the ship's naturalist, he made some tantalising observations of the biological world. On the Galápagos archipelago, 1,000 kilometres off the west coast of South America, the finches on different islands had different-shaped beaks. In all cases, the beaks were perfectly shaped for exploiting the nuts available locally: short, stubby beaks for cracking open big nuts, slender beaks for less formidable seeds.

An explanation began to form in Darwin's mind when he also noticed that the birds and animals on the Galápagos were but slight variants of those common on the mainland of South America. The Galápagos, it seemed, had been colonised by creatures from the nearby continent. Some birds and animals from South America that could easily have made a living on the Galápagos were conspicuous by their absence. Only a small subset had made it across the ocean barrier on winds or mats of floating vegetation. It had been these hardy creatures that had radiated to fill all the empty niches – a single type of finch spreading to all islands and evolving beaks best suited to exploit the seeds found locally.

Darwin was now in possession of new and important clues about evolution. But he did not know what was driving the changes in species – what was pushing each to an apparent perfect fit with its environment. Back in England in 1836, and still only twenty-seven, he sat down at his desk, laid out the facts he had collected before him, and began to think.

Darwin was aware of one common way that creatures change their forms over the generations: by deliberate breeding. Plants and domestic animals inherit physical traits from their parents, and these can be enhanced. To create a flock of sheep with the thickest-possible woolly coats, for instance, breeders select sheep with the thick coats, mate them together, and repeat the process, generation after generation.

But, whereas humans select for traits they desire in an animal or plant, nature appears to select for traits that maximise an organism's chance of survival in its environment. Such natural selection might not be as fast as the artificial selection of human breeders but it is just as effective.

After Darwin had spent eighteen months of intense concentration on the problem, a light went on in his mind. He suddenly saw the elusive mechanism of natural selection. And it was breathtakingly simple.

One of the striking things about the natural world is how profligate organisms are. Invariably, animals give birth to large litters of young. Plants produce vast quantities of seeds. But there is simply not enough food in the world to support so many young. Inevitably, therefore, most creatures starve to death. Crucially, Darwin realised, the only ones who survive to reproduce are those with traits that best enable them to make a living in their environment.[1] And these traits are inherited by the next generation. So, as time goes by, the prevalence of beneficial traits in a population increases at the expense of traits that do not confer survivability.

This was it: the missing piece of the jigsaw. Evolution by *natural selection*. 'How extremely stupid not to have thought of that,' said Darwin's friend and champion, Thomas Huxley. But of course Darwin had to see past the dizzying complexity of the natural world to the mechanism ticking at its heart and quietly generating its complexity. And that was no mean feat.

Richard Dawkins has called evolution by natural selection the greatest idea in the history of science. And it certainly has phenomenal explanatory power. Modern biology is literally the story of evolution by natural selection. 'Nothing in biology makes sense except in the light of evolution,' wrote Theodosius Dobzhansky in 1937.

According to his biographers, Darwin made no effort to publicise his idea, realising full well that it flew in the face of the Church's teaching that God created all living creatures in their final form. Only in 1858 – after twenty years of sitting on his explosive idea – was he galvanised into action. A letter arrived from a man called Alfred Russel Wallace, who, while observing nature in Indonesia and Malaysia, had hit on the exact same unifying idea of evolution by natural selection.[2] Stunned, Darwin locked himself in his study and began writing furiously.

Darwin's epochal work, published in 1859, is universally known as *The Origin of Species*, though it says essentially nothing about the ultimate origin of life, which to this day remains a deep mystery. More apposite is the book's full, though considerably more long-winded, title: *On the origin of species by means of natural selection, or the preservation of favoured races in the struggle for life*.[3]

According to Darwin, all life on Earth today has evolved over aeons of time from a common ancestral organism by the process of natural selection. The idea conflicted not only with the biblical account of creation as a one-off event but with the Church's claim that humans were, uniquely, forged in the image of God. According to Darwin, humans were neither at the pinnacle of creation nor special in any other way. They were just another animal.

Just as, in the sixteenth century, the Polish astronomer Nicolaus Copernicus showed that the Earth was not at the centre of things and occupied no special place in the cosmos, Darwin showed that humans were not at the centre of things and occupied no special place in the living world.[4]

Darwin was courageous to present a theory that flew in the face of entrenched religious orthodoxy. But he was also very honest about the theory's shortcomings, freely admitting it was incomplete. He asked people instead to judge the idea on its broad claims, which he was sure were correct, and not on the fine details, which he did not possess but which he was certain future generations of biologists would fill in.

Two things stood out as glaring omissions. The first was the mechanism of variation. People clearly inherit traits from both their mother and their father: it is possible to see a mother's red hair in a child or a father's square jaw. But what causes the appearance of new traits from which natural selection, well, *selects*?

The second thing missing from Darwin's theory was the mechanism of inheritance. Darwin initially thought that information about traits was carried from generation to generation when some kind of fluid from each parent intermingled. However, just as red and yellow paint mix together to make orange paint, while losing red and yellow for ever, combining such biological fluids should blend together traits, losing some for ever. We should see people with eyes only a blend of blue and brown and never people with undiluted blue or brown eyes, something that flatly contradicts reality. Over time, the blending of such biological fluids should cause all creatures in a population to become similar, drastically reducing the variation needed for the operation of natural selection. When Darwin realised this flaw in his fluid idea, he was deeply depressed.

The mechanism of inheritance and variation

It was a monk called Gregor Mendel in Brno, in what is now the Czech Republic, who was the first to glimpse the elusive mechanism of inheritance. Between 1856 and 1863, Mendel bred together varieties of pea plants in their tens of thousands and listed a number of traits that were inherited in their entirety. For instance, when Mendel bred pea plants with purple flowers with ones with white flowers, the result was not pea plants with a pinkish flower but a certain predictable fraction of white pea plants and a certain predictable fraction of purple pea plants. Characteristics are inherited equally, one from each parent, with some traits more dominant than others, Mendel found. Crucially, however, they are inherited as *particles* that can never be subdivided, not as a fluid that can be blended. Mendel, though he did not know it, had discovered what we now call genes.

Mendel published his findings in *Proceedings of the Natural History in Brünn* in 1866. But the journal was so local and obscure that his work was not widely recognised until the twentieth century. There is a story, often repeated, that, of the 115 copies of Mendel's pea paper, one found its way to Darwin himself. It was discovered in his library after his death, sealed and unread. It would have been a terrible tragedy if true. However, the story is mere romantic myth. Darwin had no work by Mendel in his vast collection. The two biological geniuses, each of whom possessed a crucial jigsaw piece the other lacked, missed each other not by a hair's breadth but by a significant span of space and time.

Mendel's work was rediscovered only in 1900, long after Darwin's death. Shortly afterwards, the American biologist Thomas Hunt Morgan began breeding together fruit flies. He observed that they inherited characteristics in a pattern very similar to Mendel's pea plants. He even established that the physical elements responsible for inherited traits – genes – lay on tiny stringy structures called chromosomes. It was the birth of a new science: genetics.

The full picture of inheritance was filled out only in the late twentieth century. The building blocks of all life are cells, tiny bags of chemicals, whirring with chemical nanomachinery.[5] In the centre of every cell is a mini cell, or nucleus. And, in each nucleus, chromosomes made of DNA.

DNA is a molecule the shape of two spiral staircases intertwined. The core, or backbone, of this double helix is made of a sequence of just four molecules, or bases – adenine (A), guanine (G), cytosine (C) and thymine (T) – which are joined in pairs. *A*, *G*, *C* and *T* are the four letters of the genetic code.[6] Each triplet of bases codes for a particular amino acid. And amino acids are the building blocks of proteins, miraculous molecules that can carry out all manner of biological tasks, from speeding up the chemical reactions of life to detecting sunlight in your eye to providing the scaffolding that keeps your body rigid enough not to collapse into a puddle of jelly and water.

A stretch of DNA that encodes a protein is called a gene. And herein lies the connection with Mendel. The traits he identified that were inherited were associated with genes. A particular gene, for instance, makes a protein that influences the development of a pea to be wrinkly or smooth.

There are about 3 billion letters in a strand of human DNA, accounting for about 23,000 genes. This seems a woefully inadequate number to create a human being, and biologists were truly shocked that there were not more. But they have had no choice but to live with it – 23,000 genes are all there are.

Some of the genes are involved in controlling other genes. They switch off or switch on their ability to make, or express, proteins at various times in a developing embryo. And they do this depending on factors such as the concentration of a particular chemical in the cell.[7] Such control genes cause different sections of DNA to be read in different types of cell, explaining how, despite every cell in a human being containing a copy of exactly the same DNA, some cells develop as blood cells, others as liver cells or brain cells, and so on.

But DNA explains not only the mechanism of inheritance but the mechanism of variation too. If an offspring is to inherit traits from its parents, their DNA must be copied. With a whopping 3 billion letters to reproduce faithfully in the case of human DNA, the amazing thing is how good the copying process is.[8] But it is not perfect. A mistake is made about once every 1 billion base pairs. Sometimes a letter is not copied correctly. Or a sequence of DNA is deleted or duplicated. There are a myriad possible transcription errors. In addition, changes in genes can be caused by cancer-causing chemicals, viruses, ultraviolet light and nuclear radiation.

The upshot is that over time *genes gradually change*.

There is a lot of redundancy built into DNA to minimise copying mistakes, so many of the individual changes make little difference – the protein encoded by the gene still works. Some changes are harmful, causing inherited diseases such as cystic fibrosis. But, very occasionally, a change in DNA turns out to make a beneficial change to an organism – for instance, conferring on it an increased resistance to malaria. Of course, the ultimate arbiter of what is *beneficial* to an organism is its environment. A change in a gene that results in a thick, warm coat is beneficial to an animal living in a world plunging into an ice age but not to one living in a tropical world.

It is worth pointing out that changes, or mutations, occur in the DNA of all organisms. But, whereas simple organisms such as bacteria merely create copies, or clones, of themselves when they reproduce, other creatures have sex, producing offspring with half their genes from each parent. Such a composite of different traits passed down the maternal and paternal line greatly boosts the novel gene combinations available for natural selection.[9]

Mutations explain the existence of *species* – groups of animals, which, broadly speaking, cannot interbreed. Species can arise in many ways. For instance, a geographical barrier such as a river or mountain range might split a population in two. Or, as in the case of the Galápagos, an ocean might divide creatures from their cousins on the mainland. Separated in this way and subjected to different survival pressures, the DNA of each group accumulates different mutations, so the populations gradually diverge. Eventually, the two groups can no longer interbreed.

There could be many reasons for this. It could be that a mixture of their genes simply does not lead to a working organism, in much the same way that putting a motorbike engine in a Rolls-Royce does not create a viable car. Or it could be that members of one group hang out on a particular type of fruit, waiting for a mate, whereas members of another group prefer another type of fruit entirely; though they could easily mate, they miss each other like ships in the night. In the case of insects, which have complex genitalia, two groups might no longer interbreed because one develops sex organs that, like a skeleton key and a Yale lock, physically do not fit each other.

Whatever the reasons for groups of creatures diverging from each other, natural selection has populated the world with a myriad distinct species, each with as little ability to breed with each other as humans and oak trees.

The explanatory power of Darwin's theory

Darwin's theory explains so many aspects of the world. For instance, it explains why life on Earth is so staggeringly diverse, boasting more than 5 million living species. It also explains why we share around 99 per cent of our DNA with chimpanzees – and even a third with mushrooms. This is exactly what would be expected if we evolved from a common ancestor. Since changes in genes accumulate over time, the DNA differences reflect the fact that the common ancestor of humans and chimpanzees lived relatively recently whereas the common ancestor of humans and mushrooms lived in the very remote past.

Arguably, the most remarkable DNA sequence on Earth is GTG CCA GCA GCC GCG GTA ATT CCA GCT CCA ATA GCG TAT ATT AAA GTT GCT GCA GTT AAA AAG.[10] It is present in every single living organism – even in organisms not technically classed as alive such as giant mimi-viruses. The reason the sequence is so widespread is that it existed in the common ancestor of all life. Carrying out a crucial process, it has remained unchanged for 3 billion years: the oldest fossil in your body.

Darwin's theory also explains why our antibiotics become less and less effective with time. Initially, they may kill the over-whelming majority of bacteria infecting a person. However, genetic variation within a population of bacteria ensures that some, inevitably, will survive to reproduce. Each successive generation will, therefore, contain a higher proportion of antibiotic-resistant bacteria, until eventually the antibiotic is next to useless. 'Evolution is . . . an infinitely long and tedious biologic game, with only the winners staying at the table,' says Lewis Thomas.[11]

Most of all, however, Darwin's theory explains the illusion of design – why organisms are so perfectly suited to their environments. The reason a finch on an island in the Galápagos has a beak perfect for cracking open the nuts it lives on is because its ancestors prospered, leaving more offspring than did finches with less effective beaks. The shape of a beak turns out to be controlled by a single gene, slight variants of which express different proteins in the growing jaw of a finch embryo.

The remarkable thing is that such an exquisite match between organism and environment is achieved without a designer. But, then, the natural process identified by Darwin is not random. 'Mutation is random,' says Richard Dawkins. 'But natural selection is the very *opposite* of random.'[12] It preferentially culls all the variants except those that confer on their host the ability to survive to reproduction. Incrementally, generation by generation, it accumulates advantageous changes, slowly but surely assembling machines far more exquisite and complex than any designed by humans. 'The whole trend of life, the whole process of building up more and more diverse and complex structures, which we call evolution, is the very opposite of that which we might expect from the laws of chance,' wrote American biologist Gilbert Newton Lewis.[13]

But evolution by natural selection has its limits. The only organisms that can arise are those that are the result of a long string of advantageous changes. 'Evolution walks backwards into the future,' says British biologist Steve Jones. 'It doesn't know what's coming.'[14] This has led some people to claim that Darwin's theory cannot explain the existence of complex organs such as the eye, which consists of multiple components. Until all components are in place – a lens, a light-detecting surface, and so on –

goes the argument, no advantage is conferred on an organism. What use is 50 per cent of an eye? Or 5 per cent of one?

However, it turns out that all the steps along the road to the eye were indeed advantageous. Examples of primitive eyes can be seen throughout the animal kingdom. Some creatures have only a patch of light-sensitive cells for sensing which way is up and which down. Others, like the pit viper, have light-sensitive – actually, heat-sensitive – cells at the bottom of a pit in their skin, so their 'sight' has a directional capability. From this, it is a short step to close over the pit with a transparent protein, creating a lens that can focus an image of an object.

In addition to having no foresight, evolution by natural selection does not necessarily result in more complex forms. It *can*, but it does not always do so. After the advent of the first cell, there really was nowhere to go but *up* in terms of size and complexity. But, as soon as larger creatures evolved, it was possible to evolve back down to simpler forms. This can be seen in the case of parasites, which live off their more complex hosts.

Darwin's theory of evolution by natural selection – Dawkins's 'greatest idea in the history of science' – has passed every test. 'It could so easily be disproved if just a single fossil turned up in the wrong date order,' wrote Dawkins.[15] All it would take would be the discovery of a rabbit in the pre-Cambrian period 500 million years ago. As yet, this has not happened.

4:

THE BIG BANG OF SEX

Sex

I admit, I have a tremendous sex drive. My boyfriend
lives forty miles away.

PHYLLIS DILLER

Sex is a bad thing because it rumples the clothes.

JACKIE KENNEDY

In his 1986 book, *The Blind Watchmaker*, British biologist Richard Dawkins paints a beautiful and evocative image. 'It is raining DNA outside,' he writes. 'On the bank of the Oxford canal at the bottom of my garden is a large willow tree, and it is pumping downy seeds into the air . . . spreading . . . DNA whose coded characters spell out specific instructions for building willow trees that will shed a new generation of downy seeds . . . It is raining instructions out there; it's raining programs; it's raining tree-growing, fluff-spreading, algorithms.'

Of course, the DNA in each of those countless fluffed-up seeds will remain nothing more than an inert coil of chemicals – a non-running computer program – unless it collides and merges with another chunk of DNA, kick-starting the creation of a new willow tree. Sex, as Dawkins so eloquently points out, is everywhere. It is what makes the world go round. Pretty much all creatures – from ants to antirrhinums, pine trees to pangolins, sunflowers to sail fish – indulge in it. Yet, to steal the words of Winston Churchill, sex is 'a riddle, wrapped in a mystery, inside an enigma'.[1]

The central mystery of sex is not hard to appreciate. Once upon a time, in a primeval pond on the newborn Earth, there arose molecules that could copy themselves.[2] Those that were most successful became the most numerous; those that were least successful were outcompeted for the necessary chemical building blocks and so disappeared. Eventually – and this undoubtedly took a mind-cringingly large number of steps, a vast amount of *pre-evolution* – a single type of molecule became pre-eminent because of its ability to build molecular machines that could exploit energy resources to promote its own reproduction. This was DNA – a necklace of genes, many of which coded for individual pieces of protein nanomachinery. 'All of today's DNA, strung through all the cells of the Earth, is simply an extension and elaboration of [the] first molecule,' said American biologist Lewis Thomas.[3]

Over billions of years, natural selection, which has seen some gene sequences outcompete others for resources and so propagate into the future while others have fallen by the wayside, has created the most amazingly elaborate vehicles for promoting genes. But that is essentially all they are. Whether fungi or fur seals, *E coli* or elephants, hydras or humans, they are *vehicles for propagating genes*. 'A hen is only an egg's way of making another egg,' as Samuel Butler put it.[4] And the most successful vehicles are those that get their genes into the next generation. Not just some of them. *All of them.*

The straightforward way for an organism to do this is simply to make a copy, or clone, of itself. Asexual reproduction is the strategy used by most simple organisms such as bacteria plus a few more complex organisms such blackberries. However, mysteriously, the large majority of multicellular organisms use an alternative reproductive strategy. They combine *half* their genes with *half* of the genes of another organism. This is of course sexual reproduction.

The obvious disadvantage of sexual reproduction is that, instead of passing 100 per cent of an organism's genes into the next generation, it transfers a mere 50 per cent. 'Sexual reproduction is analogous to a roulette game in which the player throws away half his chips at each spin,' says Dawkins.

Common sense says that a creature reproducing sexually can compete with one reproducing asexually only if it produces *twice as many offspring*. But this is very costly in terms of energy. And, in a world of cut-throat competition for food resources, energy efficiency is imperative for survival. But the cost of producing extra offspring is not the only extra cost of sex. It takes energy, after all, to find a partner with which to merge genes. Think of the willow tree in Dawkins's garden, which must create and release such a tremendous quantity of feathery seeds into the air of Oxford. 'The reproduction of mankind is a great marvel and mystery,' wrote Martin Luther. 'Had God consulted me in the matter, I should have advised him to continue the generation of the species by fashioning them of clay.'[5] The striking feature of the world, however, is that sex is ubiquitous. Not only do birds and bees do it, so do pretty much all plants, reptiles, mammals and birds. Clearly, it must have a huge evolutionary advantage not only to have survived but so evidently to have thrived. But what is that advantage? Remarkably, it is not obvious. Not obvious at all.

One hint of the possible advantage of sex comes from the variety of its resultant offspring. The offspring of an organism that reproduces asexually are not *exact copies* of that organism because the copying of DNA is never perfect. However, the variety produced by occasional copying errors, or mutations, pales into total insignificance compared with the variety created by sexual reproduction. If the genes of an organism are likened to a deck of playing cards, each offspring of an asexual organism inherits the same deck of cards, possibly with one card substituted by a wild card. However, each offspring of sexual reproduction inherits half the cards from two separate decks shuffled together. And, for each individual offspring, the two decks are shuffled together *differently*.

What this means is that the offspring of sexual reproduction are *very* different from their parents.[6] Sexual reproduction generates maximum novelty in the next generation. Conceivably, at times when the environment is stressed, such as when the climate is changing rapidly, sexual reproduction can throw up such a large range of organisms that some at least will have novel traits necessary for survival. By contrast, asexual organisms, terminally stuck in a rut, will die out. Is this enough of an advantage, though, to explain the survival of sex? Biologists are not entirely clear.

Another possible reason for the survival of sex is that it provides the means to combine in a single organism advantageous gene mutations *from two organisms*. Think of two asexual organisms each of which acquires a mutation in one of its genes that aids its survival. The two mutated genes are doomed to remain forever separate, isolated in each line. Sex, however, changes everything. It means that two good genes from two separate organisms can end up side by side on the same strand of DNA, compounding the survival chances of any offspring. This sounds like a big advantage. Unfortunately, of course, sex not only concentrates good genes in a single organism but also bad genes in a single organism. No one is quite clear whether the advantage sufficiently outweighs the disadvantage.

So what, then, is the overwhelming – yet mysteriously elusive – advantage of sex? One idea that has gained popularity – though not universal acceptance – is that sex wrong-foots potentially deadly parasites. Such creatures are the bane of all complex organisms. More than 2 billion people worldwide are infected by parasites, ranging from malarial protozoa to intestinal worms. Evolution by natural selection acts on parasites in the same way it does on all organisms. But a parasite's environment is *its host*. Consequently, its success at exploiting the resources of its environment comes at the cost of depleting the resources of the host. Parasites drain their host of life and may eventually even kill it. And this can all happen very rapidly since parasites are generally small and fleet of foot, capable of reproducing many times over during the lifetime of their host.

How can a population of host creatures possibly survive so relentless and effective an assault? The answer is by continually replacing its members by new members that are utterly novel and to which the parasite is not perfectly adapted. This is exactly what sex accomplishes, claimed American biologist Lee Van Valen in 1973.[7]

Yes, parasites can change rapidly. But a host population can survive if it can change *even more rapidly*, said Van Valen. In *Through the Looking-Glass*, Lewis Carroll's 1871 sequel to *Alice in Wonderland*, Alice is running alongside the Red Queen but is completely baffled that she appears to be making no discernible progress.

> 'In our country,' said Alice, still panting a little, 'you'd generally get to somewhere else – if you run very fast for a long time, as we've been doing.'
>
> 'A slow sort of country!' said the Queen. 'Now, here, you see, it takes *all the running you can do, to keep in the same place*.'

For this reason, Van Valen's parasite explanation for sex has become known as the Red Queen Hypothesis.[8]

In 2011, biologists in the US tested the idea in a controlled laboratory environment.[9] They genetically manipulated the mating system of the roundworm *Caenorhabditis elegans* so that different populations could reproduce either asexually, by fertilising their own eggs, or sexually, by mating with male worms. They then infected *C. elegans* with the pathogenic bacteria *Serratia marcescens*. The bacteria rapidly drove extinct the self-fertilising population of *C. elegans*. However, this was not the case for the sexually reproducing population. It was able to outpace its co-evolving parasites – continually *running faster* – appearing to confirm the Red Queen Hypothesis. Sex is a weapon against parasites.

The mechanics of sex

Sex involves the combination and shuffling of genes from two organisms to create an entirely novel organism. The devil, however, is in the detail. And the detail is both subtle and complex.

To appreciate it, it is first necessary to know some background. If the DNA in a single one of your cells was arranged into one straight piece, it would stretch right the way from your head to your toe. Packing all that DNA into a tiny cell, invisible to the human eye, is therefore a biological challenge. A cell achieves this impressive feat by packaging the DNA into shorter stretches known as chromosomes, so called because they were first revealed with the aid of coloured, or chromatic, dyes.[10] Human cells have 46 chromosomes – two sets of essentially the same chromosomes.

Dogs have 78 chromosomes; horses 64; and cats and pigs 38. The number of chromosomes appears to bear little relation to an organism's complexity. The Adder's-tongue fern, *Ophioglossum vulgatum*, for instance, has a whopping 1,440 chromosomes, the largest number of any living thing.[11]

Back to humans. Recall that, every day, your body creates about 300 billion new cells – more than there are stars in our Milky Way Galaxy.[12] In this process, known as mitosis, a cell first creates a copy of all 46 of its chromosomes. That makes a total of 92. Then, the cell splits into two 'daughter' cells, each with 46 chromosomes, exactly like the original.

Sex is the opposite process. Instead of splitting one cell into two, two cells – one from each parent – are fused into one. However, if the final cell is to have the correct complement of 46 chromosomes, the cells from each parent – known as sex cells, or gametes – must each contain only 23 chromosomes, or *half the normal number*.

The creation of sex cells by both males and females, therefore, requires a process quite distinct from mitosis. In meiosis, as in mitosis, a cell first makes a copy of all 46 chromosomes, to make a total of 92. But then it splits *not once but twice*. The end result is four gametes, each of which contains 23 chromosomes.

Incidentally, some shuffling of the genes is carried out during meiosis, so that each of the gametes is genetically different from its parent. This shuffling might once upon a time have been an accident – a result of the complex manoeuvring of chromosomes during meiosis – but it might have become frozen because creating offspring with the maximum amount of genetic novelty has survival value. And this shuffling of genes to create variety is *even before* sex cells fuse to generate yet more variety.

The gametes from each parent could, of course, be the same size. And this is true for some organisms. But very often one is far bigger than the other because it contains the fuel and protein machinery to drive development once fusion occurs. For biologists, the essential difference between the sexes lies in the gametes. Organisms that produce large gametes that cannot move about – known as eggs, or ova – are female – while organisms that produce small gametes that can move about – known as sperm – are male. All other things that are usually associated with the difference between the sexes – penises, vaginas, breasts and beards – are ultimately just consequences of the differences between sperm and egg.

Biologists believe that the first sexually reproducing organisms produced gametes of the same size. This is an interesting observation. It means that *sex came before sexes*.

Now, finally, we come to the fusion of two gametes – one from each parent – which is the central act of sex. Here, the two gametes, each with 23 chromosomes, combine to make a single cell, known as a zygote. Subsequently, the zygote will split, again and again, by normal mitosis, to make the 100 trillion or so cells that compose an adult human being.

Clearly, the zygote contains 23 chromosomes from the mother and 23 chromosomes from the father.[13] At a gross level, therefore, each one of your cells has two copies of *exactly the same genes*. After all, men and women are genetically more similar than, for instance, men and chimpanzees – and, recall, chimpanzees share 98 to 99 per cent of their DNA with humans.[14]

But, although your mother and father contributed the same genes to you, they may have contributed *different versions* of those genes, due to random mutations in each of their family lines. And these variants, known as alleles, can make all the difference. For instance, there is a gene that determines hair colour. The copy from your mother might, for instance, be a variant that makes you a redhead or it might be a variant that makes you a brunette. Which version of the two genes is expressed in you depends on which gene is dominant and which recessive.

There could many reasons why a copy, or allele, of a gene is dominant or recessive. It all depends on the particular gene. Each allele – one from your mother and one from your father – will make a slightly different protein. But some proteins win out over their fellows. In the simplest situation, one allele makes a *broken protein*. Since the broken protein does nothing, the working protein is dominant. A good example of a recessive allele is red hair. There is a protein called MC1R whose usual job is to get rid of red pigment. When it is not working, therefore, there is a build-up of red pigment and a person ends up with red hair.

By inheriting versions of each gene either from your mother or from your father, you inherit some characteristics from your mother and some from your father. The precise mix is random. This is how sex maximises the novelty in offspring.

Actually, it is not quite true that you have two identical sets of 23 chromosomes. In fact, you have two identical sets of *only 22*. The chromosomes in the 23rd pair differ between males and females. It works likes this. Chromosomes tend to have a characteristic 'X' shape. However, the 23rd might have a 'Y' shape. Two copies of the X chromosome make a female while an X plus a Y makes a male.[15]

All human embryos develop in exactly the same way in the beginning. However, after forty days, a gene on the male's Y chromosome called the Sex-Determining Region of the Y chromosome, or SRY, becomes active. It contains the instructions for making testosterone, which converts the gonad cells of an embryo into testes, which in turn trigger the development of male sexual organs. If the expression of SRY is blocked, however, the embryo's gonad cells become ovaries, which trigger the development of female sexual organs. Differences in hormones between the sexes cause as many as one in six mammalian genes to express their proteins preferentially in one sex rather than the other.

Males are the product of testosterone. They are females with an extra gene. And every male on Earth – even the most macho – was in touch with his feminine side for the first forty days of existence.

The big bang of sex

Since most simple organisms are asexual and the first organisms on Earth were single cells, most biologists believe that the earliest life forms were asexual. This is a hugely simpler means of proliferating than sexual reproduction. So how in the world did sex ever arise?[16]

Nature tends to adapt to new tasks things it evolved for entirely different purposes. Glutamate, for instance, one of the most important neurotransmitters in the human brain, was used by the very first bacteria for signalling almost 4 billion years ago.[17] Well, sex is no different. The basic components – the fusion of two cells, the mixing of their genes and the separation of those cells – arose for other purposes and then were co-opted for the purpose of sexual reproduction.

A fundamental process was the swallowing of one simple cell by another to create a complex cell, or eukaryote about 1.8 billion years ago.[18] This involved a multitude of changes inside a cell. For instance, the swallowee's membrane was replaced by a different type of membrane to permit it to become a cellular organelle. The exact details are not important. The point is that such adaptions made it possible for *one cell to merge with another*.

At some stage in the mists of time – and all that can be done is to speculate plausibly about this – two similar eukaryotic cells bumped into each other and *accidentally fused*. Now, some cells are known to shift to a dormant state, barely ticking over, when times are tough – for instance, during a drought. At such a time, a cell consisting of two cells fused together might have a survival advantage. After all, two cells will have *pooled their resources*. And this may not be the only advantage of the fused cell. The tough time may be tough enough actually to damage the cell's DNA. But, since the cell has two copies of its genes, it has the ability to compare the two copies and *correct any errors*.

When the good times return, a cell with only one copy of its genes will have an advantage once again. After all, with less DNA to copy, it will be able to reproduce more quickly and proliferate. This may therefore have driven the evolution of meiosis, the means of creating cells with only one copy of their genes. If this seems implausible, there are indeed single-celled organisms today that react to extreme changes in their environment by switching back and forth between a state with one copy of their genes – known as haploid – and one with two copies – known as diploid.

So much for how cells came to fuse and then unfuse in the process of meiosis. How did the DNA of two cells intermingle to create the genetic variety so central to sex? It turns out that this happens naturally in the process of repairing damaged DNA. When a cell detects a difference between the two complementary strands of DNA on a chromosome, it has no idea which strand is error-free. It therefore has no choice but simply to excise the region from both strands of DNA. This leaves a gap, which the cell fills by copying the sequence present at the same region on the matched chromosome.

All this happens when the two chromosomes are very close together. And, crucially, in the complex dance – which involves cutting up bits of DNA physically and touching them together – bits of DNA get swapped around. This process, known as crossover, ensures that, when the meiosis creates new cells, each is different from its parent. It is a happy accident that might have become frozen because natural selection favours organisms whose offspring are novel and varied.

So, it seems, sex was a simple accident that evolved into a survival strategy. It made use of pre-existing genes. Nature was left with a mechanism that unintentionally mixed DNA, greatly boosting genetic variation, causing the rate of evolution to explode.

But, of course, there is a lot more than this to sex between complex organisms such as human beings. How did *that* evolve? Nobody knows the precise details. However, it is possible to speculate on the steps along the road. First there was the evolution of cells that could fuse together and undergo meiosis. This was the origin – the big bang – of sex. Next came the evolution of sexes. Rather than a single type of cell, there arose two kinds: male and female.[19] At first, the two types were able to fuse together in all possible combinations: male–male, female–male and female–female. However, the combining of different types, or outbreeding, creates more genetic variation among offspring, which has survival advantages. Eventually, therefore, a system of sex evolved in which the only combination of cells that was viable was male–female.

In the beginning, all the cells of a sexually reproducing organism were capable of doing the deed. However, the next step in the evolution of sex was the advent of multicellular organisms in which sexual reproduction was down to *only one type of specialised cell*. One of the two types of gamete, known as sperm, evolved the ability to swim about, boosting its chance of finding the second type, known as the egg. But this was not the end of the specialisation. Eventually, the production of gametes was confined to only one type of the tissue: namely, the gonads.

Nobody knows how long all this took. But, evolution by natural selection, when it gets a good idea, runs with it. In the oceans, where life began, the sexes evolved coordinated behaviour, releasing eggs and sperms simultaneously into the water in order to maximise the chance of their union. Such a strategy was impossible, however, after animals moved onto the land. Instead, internal fertilisation became advantageous. Matched genitals evolved so that males could penetrate females and fertilise them. 'Sexual intercourse began / In nineteen sixty-three / (which was rather late for me),' wrote the poet Philip Larkin.[20] But, actually, it was rather earlier than that. Finally, to protect a developing embryo better, females evolved a womb, or uterus, in which an embryo could develop in relative safety.

The road to modern sex has been a long one but at least the major milestones along that road appear clear. Nevertheless, sex very much remains that 'riddle wrapped in a mystery inside an enigma'. And this is evident even when looking around at the human world today.

Other sex mysteries

Take homosexuality, defined as sex between a same-sex couple. Since the only way for genes and the characteristics they encode to propagate down the generations is through sex between a male and female, genes that contribute to homosexuality should, by rights, become rapidly extinct. 'We are machines built by DNA whose purpose is to make more copies of the same DNA,' says Dawkins. 'It is every living object's sole reason for living.'[21]

Yet the frequency of homosexuality is thought to be constant across cultures at about 3 per cent in men and 2 per cent in women. How can this be?

One obvious possibility is that homosexuality has no genetic component – that there is no gene or genes that determine homosexuality. In fact, Dawkins's basic 'selfish gene' idea has been increasingly tempered by the realisation that the environment plays a role in the expression of genes. According to the field of epigenetics, cells read DNA more like a script to be interpreted – depending on, for instance, environmental chemicals – than as a super-strict blueprint. 'My mother made me a homosexual,' goes the joke. To which someone replies, 'If I give her the wool, will she make me one too?'

Another possibility is that homosexuality has a genetic component that, though it is not beneficial in promoting the cause of selfish genes, comes along with a gene that is. This is not uncommon. For instance, there is a particular gene that gives people immunity to malaria. But if, instead of having one copy of the gene, a person has *two copies* – one from each parent – they get sickle cell anaemia, in which blood cells become flattened and block capillaries. Sickle cell anaemia – a cripplingly painful disease – persists because, in most people, the gene that causes it has a beneficial effect and boosts their chance of survival.

Of course, homosexuality might persist because homosexuals *do* get their genes into the next generation. Although there is a tendency to pigeon-hole sexuality, in fact there is a whole spectrum, ranging from 100 per cent heterosexuality through bisexuality to 100 per cent homosexuality. 'Sexuality is as wide as the sea,' said English film-maker Derek Jarman. People may not be totally homosexual – or might be homosexual only at certain times in their lives. This would mean that homosexuals do sire enough children – at least to make sure their genes persist through the generations, and that homosexuality persists from generation to generation.

But there is a possibly more plausible way that homosexuals could get their genes into the future. If they help in the rearing of children who are genetically related to them – perhaps the offspring of a brother or sister – they will actually be acting selfishly to ensure their genes propagate into the future. This is similar to the argument often employed to make sense of another great mystery of biology: altruism. Why do individuals do things that ensure the survival of others *at the expense of their own survival*? Again, the argument goes that people are more likely to do that to people who are genetically related to them – that is, close family members.

And this argument might help explain yet another major sex mystery: the menopause. Remarkably, humans are one of only three species known to experience a shutdown of their reproductive potential before they die. The others are killer whales and short-finned pilot whales. (You have to pity female short-finned pilot whales – not only do they suffer the menopause but they have only *short fins* to fan themselves with if they get a hot flush.)

The menopause occurs when a female is depleted of eggs. One is released from one of her ovaries every monthly cycle. But the total number of eggs is fixed at birth at about 400. They are therefore typically exhausted when a woman is around fifty.

By the way, since a woman's eggs are present in her ovaries when she was an embryo in her mother's womb, there is a sense in which you began your life not inside your mother but *inside your grandmother*.

What is so peculiar about a woman's reproductive potential shutting down before death is that the ability of a female to 'get one more in' – even late in life – would appear always to be advantageous when the name of the game is to produce as many offspring as possible. Why not have more than 400 eggs? Why not have enough to last a lifetime?

But maybe there are other factors that come into play. Certainly, later in life, childbearing is more risky for a woman and the chance of a child inheriting genetic defects is higher. Add to this the fact that successfully rearing a child to adulthood takes a large amount of energy. Not only might an older woman lack this energy but she is more likely to die while still rearing a child.

Perhaps, by switching off her ability to reproduce, a woman makes herself available to help her *children rear children*. Not only will this enhance the chance of the grandchildren sur-viving; since her children are, well, *her children*, it will boost the chance *of her own genes* propagating into the next generation. It is all costs and benefits. The cost of her own pregnancy and the subsequent child rearing set against the benefit of helping to rear a grandchild. Perhaps the latter wins out. Grandmothers do the unselfish right thing, goes the argument – out of selfishness!

5:

MATTER WITH CURIOSITY

The Brain

I am a brain, Watson. The rest of me is a mere appendix.

SHERLOCK HOLMES[1]

Brain against brute force – and brain came out on the top – as it's bound to do.

TOAD OF TOAD HALL[2]

One of the most profound questions in science is: why is the Universe constructed in such a way that it acquires the ability to become curious about itself? The question presupposes the existence of an objective Universe *out there*. Yet everything we know about reality, including our model of the Universe, is a construct of the human brain. 'The brain', as poet Emily Dickinson wrote, 'is wider than the sky.' Before we can truly address any of the really deep questions about the Universe, we first need to understand the filter through which we perceive that Universe.

Captain James T. Kirk of the starship *Enterprise* called space 'the final frontier'. But he was mistaken. It is not space that is the final frontier. It is the human brain: the ultimate piece of 'matter with curiosity'.

Our brain – 'the apparatus with which we think that we think'[3] – processes information from our senses, using it to update its internal model of the world. It then decides, on the basis of that information, what action to take. The brain is responsible for art and science and language and laughter and moral judgements and rational thought, not to mention personality, memories, movements and how we sense the world. 'It is in the brain that the poppy is red, that the apple is odorous, that the skylark sings,' wrote Oscar Wilde.[4]

Not bad for a chunk of unprepossessing matter with the consistency of cold porridge. The question is: how did something as complex and amazing come about? The answer is inextricably bound up with the origin of the nervous system – and with the harnessing of lightning.

The evolution of electricity

In the beginning, there were simple bacteria – microscopic bags of gloop with the complexity of small cities. They faced a serious problem: how to orchestrate their internal 'factories' to make the micro-machinery of life – the Swiss-army-knife molecules known as proteins. The solution they hit on was to release molecules such as glutamate, which diffused throughout their liquid interiors. When such a chemical messenger docked with a molecular receptor – fitting into a cavity like a key into a lock – it triggered the cascade of chemical reactions needed to make a protein.

After almost 3 billion years stalled at the single-cell stage, life made the giant leap to multicellular organisms. But it continued to use its ancient, tried-and-tested system of internal communication. Take sponges, for instance. These colonies of cells pulse in synchrony in order to pump food-laden water through channels in their bodies. Sponge cells achieve this feat of coordination by detecting chemical messengers such as glutamate, which are released by other sponge cells. It is nothing more than what happens inside a single bacterium writ large. If it ain't broke, don't change it, as far as nature is concerned.[5]

The chemical messengers of a sponge take many seconds to diffuse to all of its cells and trigger a response. This is acceptable for a creature living in surroundings that are constant and predictable. However, in a rapidly changing environment, where a quick response to threats is essential for survival, a faster method of internal communication is imperative. Such a means is provided by electricity.

Remarkably, electricity is as ancient a feature of cells as chemical messengers. Cellular membranes are leaky and prone to let through dangerous charged atoms such as the sodium in salt.[6] In order to survive, bacteria needed a way to pump out such ions. They solved the problem with the aid of tunnel-like proteins called ion channels, which span the cell membrane and can open and shut to expel ions. But, inevitably, pumping ions through such a channel creates an imbalance of electric charge between the inside and the outside of the cell. It is this voltage difference that provides a cell with a nifty communication opportunity.[7]

To send a super-fast signal, a cell needs only to manipulate the voltage across its membrane, which it can do simply by pumping ions rapidly through an ion channel. This causes an abrupt change in the voltage across the membrane, which has a knock-on effect on the next ion channel, and the next, and so on. Like a microscopic Mexican wave, an electrical signal propagates along the membrane, thousands of times faster than any chemical messenger, *literally at lightning speed*.

Of course, a communication system based on electricity – a true *cellular telephone system* – needs a means not only of transmitting a signal but a means of detecting it at its destination and doing something useful with it. Cells have this covered too. A type of channel known as a voltage-gated ion channel can open in response to an electrical signal, allowing ions such as calcium to pass through the membrane. These then trigger a cascade of cellular processes, effectively turning the incoming electrical signal back into a bog-standard chemical messenger, which can do something useful such as trigger the building of a protein.

Voltage-gated ion channels, just like regular ion channels, are present in bacteria. Cells that use them for internal communication simply borrowed them and adapted them to the new and specialised task.

An internal cellular telephone system was in existence even before the first multicellular animals. In fact, it can be seen in action in a water-living, single-celled creature called *Paramecium*. When *Paramecium* is swimming along and bumps into an obstacle, a voltage is created across its membrane. This causes a Mexican wave of ions to pulse around its body. Lightning fast, the wave reaches hair-like extensions on the surface of the cell, which, when they ripple in synchrony, can propel the cell. Instantly, these cilia reverse their beating, causing *Paramecium* to back away from the obstacle.

A useful trick for a single-celled creature such as *Paramecium* turns out to be indispensable for a multicellular organism. After all, as creatures grew ever larger, it became likely that the place on their bodies where they sensed a dangerous touch was a *long way* from the place where a muscle had to be contracted in response. Sending a signal via a chemical messenger was far too slow. Long before an animal could take evasive action, it might be eaten. Electricity was the only solution. And nature responded by creating a specialised electrical cell – the nerve cell.

Nerve cells

A nerve cell has a cell body with a nucleus like a normal cell. But there the similarity ends. One side of the cell is extended like a long, thin wire, while the other side sports a number of finger-like extensions. The long, thin wire, known as an axon, *transmits* an electrical pulse to another nerve cell, whereas the finger-like extensions, known as a dendrites, *receive* electrical signals from the axons of other nerve cells.

Crucially, the axon of one nerve cell does not touch the dendrite of another. There is a gap – known as the synapse. Here, the electrical signal from the axon is converted into chemical messengers.[8] These diffuse across the gap and dock with receptors, which open ion channels and thus trigger a new electrical signal. Sound familiar? It is the very same molecular lock-and-key system that bacteria came up with almost 4 billion years ago. Life, far from discarding its ancient and sluggish communication system, mediated by chemical messengers, *integrates* it into its superfast and modern communication system, mediated by electricity.

The mediation of the electrical signal by chemical messengers is not just an unfortunate hangover from the beginning of life. It makes it possible for an almost infinite array of responses from a nerve cell. This is because there are a host of different chemical messengers, or neurotransmitters, each of which has an effect on a dendrite if and only if the dendrite possesses a receptor for it. Some trigger, or excite, an electrical current in the dendrite whereas others prevent, or inhibit, a current.

The two most important neurotransmitters in the human brain are glutamate – the fossil relic of the system of chemical messengers used by bacteria billions of years ago – and gamma-aminobutyric acid, or GABA. Virtually all communication between nerve cells, or neurons, in the brain is mediated by these two simple amino acids. Other neurotransmitters such as dopamine and acetylcholine merely moderate their action. Most drugs that affect behaviour work by blocking or mimicking a particular neurotransmitter, thus stimulating a receptor site and generating the same effect as the neurotransmitter. For example, lysergic acid diethylamide, or LSD, a mere speck of which causes dream-like psychedelic hallucinations, has a chemical structure very similar to the neurotransmitter serotonin.

Because a nerve cell has extensions capable of both *sending* and *receiving* electrical signals, it can join together with others in a network, with each nerve cell connected via its dendrites to the axons of many other nerve cells. Such a network can behave in a complex way.

Even a single nerve cell can exhibit memory. Say an electrical signal from a sense – perhaps touch – comes in along a dendrite and triggers the nerve cell to send a signal along its axon to contract a muscle. If, in addition to going to the muscle, the axon splits and part of its signal feeds back into a dendrite of the nerve cell, it triggers contraction again. And again. And again. A nerve cell can refire about every hundredth of a second. In this way, the nerve cell *remembers* the stimulus. If four nerve cells are connected together, they can exhibit complicated behaviour such as contracting a muscle to move away from either a stimulus on the left side or on the right side of an animal. This gives some hint of the complex behaviour possible if nerve cells connect together not in quartets but hundreds or thousands or even *hundreds of billions*.

The earliest nerve cells, though connected to each other, were also connected to the external world – receiving an input signal directly from senses or providing an output signal to, for instance, contract a muscle. There was no computation in between. However, at some stage in the history of life, nerve cells began to connect *only to other nerve cells*. This enabled such neurons to process the input information from the environment in new and complex ways in order to decide on an appropriate response. It was an epochal moment in the history of life. It marked the birth of the *brain*.

Brains

'Basically there are two types of animals,' says Columbian neuro-scientist Rodolfo R. Llinás. 'Animals, and animals that have no brains; they are called plants. They don't need a nervous system because they don't move actively, they don't pull up their roots and run in a forest fire! Anything that moves actively requires a nervous system; otherwise it would come to a quick death.'[9]

A neuron is often likened to a logic gate of a computer.[10] A logic gate, built from transistors, can be wired together with other logic gates to create a circuit that, for instance, adds together two numbers. But, whereas a logic gate has only two electrical inputs and spits out a signal that depends on the current flowing in those two inputs, a neuron can have 10,000 or more dendritic inputs, and spit out a signal that depends on the complex interplay of all those electrical inputs on numerous neurotransmitters and receptors at the nerve cell's synapse. So, although it is true that a neuron is the fundamental building block of a biological com-puter, just as a logic gate is the basic building block of a silicon computer, it is more than this. A neuron is a *computer in its own right*.

Brains, built of neurons, are expensive to run. The human brain accounts for a mere 2–3 per cent of the mass of an adult yet guzzles about a fifth of the body's energy when resting.[11] Having said this, the brain does all of its mega-computation on roughly 20 watts of power, the equivalent of a very dim light bulb. By comparison, a supercomputer capable of an analogous rate of computation requires 200,000 watts – it is 10,000 times less energy-efficient than the brain.

For some creatures, however, the energy expense of running a brain is simply too great. The juvenile sea squirt has a rudimentary nervous system that enables it to wander through the sea searching for a suitable rock or hunk of coral to cling to and to make its home. 'When it finds its spot and takes root, it doesn't need its brain any more,' says American cognitive scientist Daniel Dennett. 'So it eats it!'[12]

Despite this rather disturbing example of autocannibalism, the benefits of having even a simple brain usually appear to outweigh the costs. For instance, the nematode worm, *Caenorhabditis elegans*, has a brain with a mere 302 neurons – so few that its brain is completely encoded in its DNA. The nematode worm, unlike the sea squirt, does not eat its brain. It must therefore provide the worm with an important competitive advantage.[13]

The human brain weighs about three pounds and has about 100 billion neurons – by sheer coincidence, roughly the same number of stars in our Galaxy, galaxies in our Universe, and people who have ever lived. 'The human brain is the most complex object known in the Universe,' says Edward O. Wilson, '*known, that is, to itself*.'[14] According to a theory developed by American neuroscientist Paul MacLean, in the course of evolution three distinct brains have emerged, accreting one on the other. 'With modern parts atop old ones, the brain is like an iPod built around an eight-track cassette player,' says American journalist Sharon Begley.[15]

The oldest and most primitive part of our three-pound universe includes the brainstem and the cerebellum, which turn out to be the main structures in the brain of a reptile. Our 'reptilian brain' controls vital automatic functions such as body temperature, breathing, heart rate and balance. Wrapped around the reptilian brain is a structure that developed in the first mammals about 200 million years ago. The main parts of this limbic system are the hippocampus, amygdala and hypothalamus. They record memories of good and bad experiences and so are responsible for emotions. Wrapped around the limbic brain is the largest structure of all, which first became important in primates. This cerebrum, or neocortex, can overrule the knee-jerk responses of the more primitive parts of the brain. It is responsible for language, abstract thought, imagination and consciousness. It has an almost boundless ability to learn new things and it is the seat of our personality. In short, the neocortex is what makes us human.

Actually, there is one more layer wrapped around the reptilian brain, limbic system and neocortex – and that is, of course, the hard bony shell of the skull. 'Because important things go in a case, you got a plastic sleeve for your comb, a wallet for your money and a skull for your brain,' observed George Costanza in *Seinfeld*.[16] The skull is actually reinforced by three layers of protective tissue known as the meninges, in between which is a special shock-proof liquid known as cerebrovascular fluid. An infection here causes the potentially fatal inflammation known as meningitis.

The neocortex is divided into two hemispheres, connected by a bundle of nerve fibres called the corpus calossum. In effect, therefore, we have *two* brains. Usually, the left side is better at problem solving, maths and writing while the right side is creative and better at art or music. For reasons that are not completely understood, the left side of the brain controls the movement of the right side of the body and vice versa. This is why people who suffer a stroke in the left side of their brain lose movement on the right side of their body and vice versa. A stroke is usually caused by a blood clot in the brain that blocks the local blood supply, damaging or destroying nearby brain tissue.

But the wonder of the brain is not in its gross structure but in its microstructure – in its 100 billion or so neurons and 1,000 billion other support cells, which surround the neurons and their axons, providing them with energy and generally keeping them healthy.[17] However, the sheer number of neurons reveals little about the operation of the brain. 'The liver probably contains 100 million cells,' says American neuroscientist Gerald D. Fischbach. 'But 1,000 livers do not add up to a rich inner life.'[18]

The key to the brain's amazing capabilities are the *connections* between its neurons. 'All that we know, all that we are, comes from the way our neurons are connected,' says Tim Berners-Lee, inventor of the World Wide Web.[19] A single neuron may posses 10,000 or so dendrites through which it can interact with 10,000 or so other neurons. In total, the brain may contain something like 1,000 trillion connections.

The big question is: how does all this mind-bogglingly complex neuronal circuitry allow us to remember things and to learn things?

Memory and learning

The common experience of memory is that we remember things that are important to us and forget things that are no longer important to us. Of course, we all forget the occasional important thing, like where we put down a book we were reading or a shopping list scrawled on a scrap of paper. But, by and large, we remember and learn things if they are significant to us – that is, *connected to things we already know*. If you hear a new word in French and you already speak French, you are far more likely to remember it than if you do not speak French. If you know how to balance on a skateboard, you will learn how to balance on a surfboard more easily than someone who has never used a skateboard.

In addition to this, *repetition* seems to be important to remember and learn things. Babies learning to speak repeat the same words over and over. Children learn times tables by reciting them over and over again until they are finally drummed into their skulls. People learning the guitar strum the same sequence of chords, hour after hour.

None of this, of course, tells us how the brain's neuronal circuitry enables us to remember things and learn new skills. But it does hint that two crucial processes in the brain are *making connections with things we already know* and *repetition*.

The things we *already know* are encoded in the pattern of connections between the brain's 100 billion neurons, just as the knowledge of how to contract a muscle to move away from a stimulus in the four-neuron network mentioned earlier was encoded in the connection between the quartet of neurons. Nobody knows exactly how the pattern encodes complex information. Although it is perfectly possible to point to a bunch of magnetic memory domains in a computer and say, 'That is storing a 6 or the letter P', it is not yet possible to point to a bunch of interconnected neurons in the brain and say they are storing the smell of newly baked bread or the knowledge of how to balance on one leg. Nevertheless, all the evidence points to the pattern of connections between neurons being key to what we know.

The connections between neurons are made by dendrites. Dendrites are therefore synonymous with what we know. To remember something or learn a new skill, therefore, something must happen to the dendritic connections between neurons.

Imagine two neurons that are connected – the axon of the first attached to a dendrite of the second. Now imagine that the first neuron starts firing because it is receiving some stimulus – perhaps some sensory information from the outside world. Remember, the dendritic connection between the two neurons represents something we already know.

Now, if the stimulus is repetitive *and* related to what we know – and the neurotransmitters in the synaptic gap between the axon and the dendrite are primed to amplify the electrical signal if it is related – the dendrite strengthens its connection. This can happen in many ways, but one way is for the dendrite to grow a large number of spines that multiply its connection points.

Of course, two neurons connected by a single dendrite can encode only a ridiculously minimal grain of information. However, since all you know is encoded in the totality of dendritic connections in your brain, by strengthening the connections not just between pairs of neurons but the connections between large numbers of neurons, new knowledge is permanently connected to something you already know and a *memory is laid down*. 'That is what learning is,' wrote novelist Doris Lessing. 'You suddenly understand something you've understood all your life, but in a new way.'[20]

'Whenever you read a book or have a conversation, the experience causes physical changes in your brain,' says American science writer George Johnson. 'It's a little frightening to think that every time you walk away from an encounter, your brain has been altered, sometimes permanently.'[21]

By this process of strengthening connections between neurons, the network that encodes all you know continually changes. But it not only strengthens connections, it makes new connections and it loses some as well. Think of the neural network of the brain as a vast thicket. In places it is growing and in other places it is being pruned back, as connections are lost between neurons that share nothing in common. This is the process of you forgetting.

What the brain can do that nothing else in the known Universe can do is constantly rebuild and rewire itself. 'The principal activities of brains are making changes in themselves,' according to Marvin Minsky.[22]

As for learning a new skill, it is a very similar process to laying down a memory. Say riding a bike requires using certain muscles. Strengthening of the dendrites that connect to neurons that control such muscles makes it easier and faster to control them. Thus, just as a memory is encoded in a network of neurons, a skill such as riding a bike or reading a book is encoded in a network of neurons. It becomes hard-wired, automatic.

This strengthening and weakening of connections between neurons or the creation of new connections to modify the network is known as neuroplasticity. Even for me to concoct this explanation, neuroplasticity had to occur in my brain. And neuroplasticity had to occur in your brain for you to understand my explanation. (If you did not understand it, no new permanent connections were made and I have left your brain just the way it was before!)

The brain is a computer but it is a remarkable kind of computer. Whereas a silicon-based computer carries out a task according to the program fed to it by a human being, the brain has no external programmer. It is a *self-programming computer*.

A baby is born with a network of neurons and the potential to connect them in a bewilderingly large number of possible ways. The programming of the baby's brain – the growing of new connections, the strengthening of some connections and the pruning back of many more – is done by its experience of the world, the information flooding in, hour by hour, day by day, through its eyes, ears, nose and skin.

Although it is very hard to see individual neurons forging links with neighbouring neurons, it is perfectly possible to see the brain programming itself at a much coarser level. The technique of functional magnetic resonance imaging (fMRI) reveals areas of the brain that are working when a person is performing a particular task. For instance, when people have been taught to meditate, it has been possible to see new areas of their brains light up in fMRI scans – new programming. Perhaps one of the most famous examples of fMRI research is a study of London taxi drivers. Eleanor Maguire of University College, London, showed how a region of the drivers' brains – that associated with spatial awareness – was actually larger than in non-taxi drivers.

'The brain is a muscle. Use it or lose it,' seems a facile statement. But – apart from the small matter of the brain not being a muscle – the 'use it or lose it' mantra encapsulates a deep truth about the brain. Just as exercising with weights encourages physiological processes that grow more muscle cells, the processes of remembering things, learning things and so on, encourages the brain to grow more neuronal connections. And, just as not exercising causes muscles to atrophy, not exercising the brain causes it to weaken or lose altogether many of its existing neuronal connections. Even Charles Darwin, who knew nothing of neurons, realised the truth of 'use it or lose it'. 'If I had to live my life over again, I would have made a rule to read some poetry and listen to some music at least once every week,' he wrote in his autobiography. 'For perhaps the parts of my brain now atrophied would thus have been kept active through use.'

Neuroplasticity is the brain's big secret. Like natural selection in evolution and DNA in genetics, it is an idea so central to understanding the brain that, without it, nothing makes any sense. Neuroplasticity explains how new experiences constantly rewire the brain – the ultimate lump of programmable matter. It explains how the blank slate of a baby's brain becomes an adult brain. It explains how a stroke victim may recover lost faculties when the task of the afflicted neurons is taken over by neurons in an adjacent area of the brain. Rehabilitation is long and hard because the process of reprogramming is analogous to a child learning skills for the first time.

And neoplasticity persists as long as you live. Your brain will still be able to make new connections even when you are a hundred years old. A centenarian can learn to use a computer – they might not learn as fast as a child but they can do it.

Can the brain understand the brain?

'The brain boggles the mind,' says James Watson, co-discoverer of DNA.[23] It remains the last and grandest frontier in biology, the most complex thing we have yet discovered in our Universe. But we have taken the first tentative steps along the road to understanding it. Nevertheless, there is still a long way to go. But is the destination even reachable? 'If the human brain were so simple that we could understand it, we would be so simple that we couldn't,' wrote the American biologist Emerson M. Pugh.[24]

Logically, Pugh is correct. The human brain can never completely understand the human brain. It would be like suspending yourself in mid-air by yanking upwards on your shoe laces. However, the brain is not trying to understand the brain. *Many brains* are trying to understand the brain: the combined minds of international scientific community. 'All the brains are not in one head', as an Italian proverb puts it.

We are still no closer to answering the question posed at the beginning of this chapter: why is the Universe constructed in such a way that it acquires the ability to become curious about itself? But, if we understand the brain, we shall finally be able to address it. 'As long as our brain is a mystery,' said Santiago Ramón y Cajal, the father of neuroscience, 'the Universe, the reflection of the structure of the brain, will also be a mystery.'

6:

THE BILLION PER CENT ADVANTAGE

Human Evolution

Man still bears in his bodily frame the indelible stamp
of his lowly origin.
CHARLES DARWIN, *The Descent of Man*

We are just an advanced breed of monkeys on a
minor planet of a very average star.
STEPHEN HAWKING, *Der Spiegel*, 17 October 1988

Once upon a time, there was a primitive species of forest-dwelling ape. For some reason it split into two separate populations. Perhaps it became divided by a mountain range or a treeless corridor; nobody really knows the truth. However, because the two populations became subject to different survival pressures, they diverged and turned into two distinct species. One species was destined to lead to chimpanzees, the other to human beings.

The precise date of the fork in the road between the ancestors of humans and our closest living relatives is not known. But the best bet is that it happened between about 6 and 7 million years ago. This is so recent in evolutionary terms that it explains why we share an astonishing 98–99 per cent of our DNA with chimpanzees. Strikingly, however, chimpanzees do not use language, build cities, program computers or fly to the Moon. Somehow, the 1–2 per cent genetic difference has been amplified into a billion per cent advantage in the real world.

Understanding how such a minuscule difference in DNA can make such a big difference in practice involves understanding a subtlety of DNA. The popular picture is of a molecule that encodes a series of instructions, or genes, for the building of proteins that determine everything from eye colour to blood group. But there is more to DNA than this. Some genes have the ability to switch on and switch off other genes, controlling the order in which they are read out, or expressed, in a developing embryo.

Although such regulatory genes account for only a small fraction of the 1–2 per cent difference between the DNA of humans and chimpanzees, crucially they have a dominant effect on the process of development.[1]

Think of regulatory genes as the recipe and standard genes as the ingredients. Similar ingredients, when combined according to different recipes, can create very different dishes. Take eggs. Depending on how they are cooked (or not), it is possible to end up with raw eggs, soft-boiled eggs, hard-boiled eggs, pickled eggs, poached eggs, fried eggs, scrambled eggs, an omelette, and so on. In the same way, similar genes, combined according to different molecular recipes, can create animals as radically different as humans and chimpanzees.

But regulatory genes show only that it is possible for small differences in DNA to create two species as different as human beings and chimpanzees. They do not tell us *how it happened*. For that there is no substitute for the fossil record.

Two legs good . . .

Charles Darwin came up with a possible story of human evolution. Our ancestors, he claimed, first stood up on two legs. This freed their hands to make tools. Making tools required more mental power, which in turn caused their brains to grow. It is an eminently plausible story except for one thing: it is contradicted by the fossil record.

Millions of years before they left any tools *and* millions of years before they had big brains, our hominin[2] ancestors in Africa were walking on two legs. *Australopithecus anamensis*, for instance, was a hominin barely more than a metre high with an ape-sized brain. There is evidence from a fossil shinbone, or tibia, that, *before* 4 million years ago, it was walking on two legs, or bipedal, much of the time. The rest of the time, presumably, it was still hanging around in the trees. Then there is a close relative, *Australopithecus afarensis*. According to the evidence of a fossil leg bone, it was walking upright by about 3.5 million years ago.

But surely one of the most evocative discoveries in all of palaeoanthropology was made by Mary Leakey at Laetoli in Tanzania in 1976. Some time, around 3.6 million years ago, three *australopithecines* padded on two legs across a bed of freshly fallen volcanic ash. They left their fossilised footprints there for all of posterity. 'Who does not wonder what these individuals were to each other, whether they held hands or even talked, and what forgotten errand they shared in a Pliocene dawn?' says Richard Dawkins.[3]

Walking on two legs releases the hands to carry food or off-spring, to fashion tools and to brandish weapons. It also allows an ape to range further afield for food and to spot predators at a greater distance. Darwin was right in believing that bipedalism has distinct advantages. The difficulty is in explaining how it came about. The only changes perpetuated by evolution by natural selection are ones that are *immediately advantageous*.[4] However, being on two legs requires a major change in the structure of the leg bones – longer thigh bones, and a shorter and wider pelvis – and the development of a powerful bottom muscle, or *gluteus maximus*, to keep those bones upright and power a running gait. Until both of these changes have been made – which is likely to have taken many generations – there appears to be no survival advantage to being on two legs.

One intriguing possibility is that the first steps towards bipedalism were taken not on the ground but up in the trees. Gibbons and orang-utans often saunter on two legs along branches so they can reach the juicier leaves and fruit at the very tips. Our ancestors might have learned this same trick. Then, when they descended to the ground, they continued the walking habit, bounding on two legs between trees.

The naked ape

Eventually, something drove our ancestors to stay on the ground permanently, where they perfected their unusual bipedal mode of locomotion. The most likely thing is that the climate gradually became drier. The dwindling rains caused the forest habitat of our hominin ancestors to shrink and be replaced by open grassland across vast tracts of their African homeland. When other creatures adapted to this new habitat, the lure of the vast grazing herds simply became too great. First cautiously, then with gathering boldness, our ancestors ventured from the leafy shadows out into the unforgiving sun.

It was *Homo erectus*, between about 1.8 and 1.9 million years ago, that made the transition to a recognisably human body shape. Walking upright was an advantage on the exposed grasslands because it minimised the body area presented to the sun. Most likely this was the time when our ancestors lost their fur. Clothed in skin alone, *Home erectus* was able to lose heat efficiently by sweating.

Actually, we are not quite the naked apes that we at first sight appear. Modern humans actually have as many hairs per square centimetre on their bodies as chimpanzees. However, evolution has made most of this human hair too fine or light to be easily seen.

Homo erectus, with its long legs driven by a powerful bottom muscle, perfected bipedalism. Like Bruce Springsteen, it was 'born to run'. Alert for signs such as circling vultures, our ancestors might first have used their long legs to reach carcasses before the scavenging competition could get there. Later, they might have pursued game over great distances, running on and on until the prey was exhausted, a hunting strategy still used by the Bushmen of Southern Africa's Kalahari desert. Although animals such as antelopes could run much faster than *Homo erectus*, they could not sustain their running for long. Our relentless marathon-running ancestors simply wore them down. Remarkably, no other predator, not even the wolf, has comparable staying power.

Meat is a more concentrated source of energy than plants. When it became a component of the *Homo erectus* diet, it made possible the growth of the brain, an incredibly energy-hungry organ that monopolises about 20 per cent of the body's total energy.[5] A chimpanzee, on its meagre plant diet, could not power anything approaching a human-sized brain.

Whether or not *Homo erectus* killed an animal itself or located one that had died in some other way, it would have faced a serious problem at the scene: how to defend the carcass against other carnivores, at least until it could detach enough meat to carry away. The solution might have been to use tools such as rocks or clubs, perhaps even carefully fashioned spears.

A million years of boredom

The first evidence of stone tools comes from about 2.6 million years ago, the twilight years of the *australopithecines*. This is a puzzle. The received wisdom is that tools allowed our ancestors to butcher meat, making redundant their long canine teeth. With no need for a bulky jaw muscle to power such teeth, it was possible for the skull and brain to grow bigger. However, the fossil evidence does not bear this out. Rather, it shows that the teeth of our ancestors shrank in size long before the advent of tools – in fact, as far back as 4 million years ago when hominins were already walking upright.

One possibility is that tools were used far earlier than 2.6 million years ago but that they were predominantly made from branches, which were not preserved as fossils; or they were made from bones, making them indistinguishable from the bones of fossilised skeletons. Such a possibility was envisioned by science fiction writer Arthur C. Clarke in the 1960s. In a memorable scene in his *2001: A Space Odyssey*, a 'man-ape' – admittedly schooled by an ET artefact! – picks up animal bones and, with a wicked gleam in his eyes, realises he can use them to *kill*.

The first tools were cobbles that had been shattered to expose a cutting edge. Although they appeared about 2.6 million years ago, bizarrely, their design remained unchanged for about a million years. Imagine if the design of aeroplanes or computers or houses remained stagnant for *40,000 generations*. It was not until about 1.7 million years ago that more sophisticated stone hand axes appeared, with longer, more worked edges than their cobble predecessors. But history then repeated itself. Hand axes remained unchanged for even longer than shattered cobbles – an astonishing 1.4 million years, or almost *60,000 generations*.

This million years of boredom – *twice over* – is remarkable when contrasted with the rapid technological changes of the past 10,000 years. One possibility is that the shattered cobble and hand axe worked perfectly well. If it ain't broke, why change it? But just because the tools did not change does not mean that our ancestors did not. Their technological ingenuity might have been used to make tools of wood or bone, which did not survive. Even if this were not the case, our ancestors underwent profound changes – for instance, in their social interactions. Hunting and scavenging and securing prey against big predators undoubtedly required a high degree of cooperation between individuals. 'The solitary, isolated human being is really a contradiction in terms,' says Archbishop Desmond Tutu. 'We need other human beings in order to be human.'[6] Just as one bee is no bee – *Una apus nulla apes*, according to the Latin proverb – one human is no human.

Within their social groups, males and females – at least during the later stages of human evolution – were roughly the same in size. Such a similarity is relatively rare among primates but common among animals that are monogamous. If our ancestors were predominantly monogamous, which seems likely, monogamy might have come about because males fighting over females did not predispose them to cooperate in hunting. In a modern example, the affair of England captain John Terry with the ex-girlfriend of a fellow team member lost him the England captaincy. Winning football games requires cooperation on a football pitch, and the players no longer trusted Terry.

Out of Africa

Around about 1.8 million years ago, *Homo erectus* left the cradle of Africa, spreading to western Asia and then to eastern Asia and southern Europe. Fossil remains have so far been found in China, Java and Georgia. At this time a lot of sea water was tied up in ice caps so South East Asia was a much more extensive peninsula than it is today, and no boats were needed to reach places such as Java. The discovery of diminutive hobbit-like skeletons on the Indonesian island of Flores created a sensation in 2003. One possibility is that *Homo floresiensis* was a descendant of *Homo erectus*. Another is that it is a descendant of a hominin that left Africa even before *Homo erectus*.

The *Homo erectus* migration was the first of several known waves of colonisation that rippled out from Africa. The migration, like so many events in human history, might have been driven by climate change. The continent of Antarctica had long been iced over. But, when North and South America joined up, it was no longer possible for warm tropical water to flow between the Atlantic and Pacific Oceans.[7] This caused ice to build up at the North Pole, gradually cooling the planet and drying it by sucking moisture from the air. The African grasslands became deserts at times, forcing some African animals to migrate into the Middle East.

About 2 million years ago, in particular, two species of sabre-toothed cat migrated out of Africa, probably in pursuit of fleeing herds of grazing animals. Our ancestors might not only have followed the herds but the fearsome predators themselves. Sabre-toothed cats abandon their carcasses, and our tool-wielding ancestors would have been the only creatures capable of cracking open the bones and skulls to obtain the energy-rich marrowbone and brains.

There might have been a second surge out Africa about 600,000 years ago. *Homo heidelbergensis*, named from a fossil jawbone found near Heidelberg in Germany in 1907, was the ancestor of the Neanderthals and modern humans. Then, finally, about 60,000 years ago, modern humans flooded out of Africa and across the world – and, eventually, even across space to the Moon.

It is worth pointing out that, although Africa is widely considered the cradle of human evolution, it is quite possible that some of that evolution occurred outside Africa among hominins that then *returned to Africa*. At present, however, the fossil record is too coarse to reveal such fine detail.

Ice ages

The huge evolutionary changes in hominins are, as far as we know, unprecedented in any animal in the history of life on Earth. They coincided with the repeated advance and retreat of ice from the Earth's polar regions over the past 2 million years. Ice ages are generally caused by cyclical changes in the orientation of the Earth's spin axis and orbit. But what appears to have magnified these Milanković cycles[8] – named after Serbian astronomer Milutin Milanković who discovered them – over the past 2 million years is the rise of the mighty Himalayas, which changed the circulation of the air around the globe. The connecting up of North and South America also closed the tropical channel through which water could be exchanged between the Pacific and Atlantic oceans, boosting a north–south flow.

Living on a planet that repeatedly iced up, hominins were continually subjected to stress by their environment, becoming extinct in cold northern regions and surviving only in regions nearer the equator. But, unlike all other creatures on Earth, whose response to advancing ice was merely to migrate to less harsh climes or to go extinct, our human ancestors were unique in having an ability to change their behaviour in response to their changing environment. Early on in their history, being adapted for change might not have been enough for them to cope with very rapid climate change. But, later on, as their culture became more sophisticated, they adapted caves as shelters from the cold; fashioned clothing out of animal skins, and harnessed fire.

Fire

Nobody knows when and how fire was first tamed. There is disputed evidence from South Africa that it happened as long ago as 1 million years but good evidence exists only as far back as a few hundred thousand years. Probably, the first fire to be used by humans was a natural fire. Someone, on a bitterly cold night, carried a smouldering branch – ignited perhaps by lightning – into the mouth of a freezing cave. Only much later did people learn how to *make* fire. This is a difficult skill that, even today, very few people possess. It is possible that the secret of making fire was discovered and lost repeatedly whenever someone who knew the secret died.

This might throw some light on why technological progress such as improvements in tools was so slow for enormously long periods of time before there were explosions of creativity. As long as our ancestors lived in small, scattered groups, knowledge might have been gained, lost, gained and lost again, repeatedly. Only when hominin numbers swelled sufficiently was there a chance of ideas surviving, spreading and spawning new ideas.

Fire made possible cooking, arguably one of the most important developments in human history. Cooking detoxified some plant poisons, boosting the range of plants that could be eaten safely, and it killed parasites in meat. But, most importantly, cooking broke down the proteins in meat so that they were easier to digest, doing some of the work of the gut. Just as a tool is an enhancement of a limb, a cooking pot is an enhancement of the stomach. More than that, it is an *external stomach*. It means that the stomach can be smaller, and less energy-hungry. And this frees up yet more energy for the insatiable needs of an ever-growing brain.

Since *Homo erectus* had small teeth and small jaws, it is possible that as early as 1.5 million years ago it was cooking its food. The first strong evidence of cooking, however, is from Neanderthals and early modern humans about 200,000 years ago.

Neanderthals

In the icy world, our direct ancestors came up against many of their hominin cousins. It is striking that none of them has survived. Most intriguing is the case of the Neanderthals, descendants of an earlier wave of colonisation of the world. Being shorter and wider-bodied than modern humans, they were built for the cold. They made tools, buried their dead and, for a long time, appeared to be thriving. They probably had language, which is believed to have originated at least 500,000 years ago. However, their vocal regions indicate that they might have uttered a smaller range of sounds than humans and that the sounds might have been higher pitched. Ultimately, however, Neanderthals died out, their last known outposts being caves on the southern coast of the Iberian peninsula.

There has long been a suspicion that Neanderthals were wiped out by modern humans. The truth, however, might be more subtle. For instance, about 2.5 per cent of the DNA of modern humans living outside Africa is believed to be Neanderthal, indicating that there was interbreeding between the two species. It might be that our ancestors had a small 1–2 per cent competitive advantage over their cousins. Magnified over many, many generations, this could have seen them monopolise ever more territory and game.

One such advantage that humans had over Neanderthals was . . . *sewing*.

Human needles are found from about 40,000 years ago. But no Neanderthal needle has ever been found. Being able to sew enabled our ancestors to make better clothes. Better baby clothes might have ensured that, in a bitter cold snap, human newborns had a slightly higher chance of surviving than Neanderthal neonates. It could have been just enough to see humans prosper at the expense of Neanderthals.

By about 50,000 years ago, *Homo sapiens* was king. Probably, no hominin species had ever amounted to more than about 10,000 to 100,000 individuals – at the most, a million. But, by 2012, *Homo sapiens* had expanded to 7 billion individuals, filling every niche on the globe and threatening the survival of every other species on Earth.

The future for humans

Some biologists argue that evolution has now stopped for human beings. We have adapted our environment to us and no longer need to adapt to it. But this ignores the fact that most people, apart from those in the affluent Western world, face a daily battle for survival as challenging as that faced by their ancient hominin ancestors. Even in the West, the demands of an ever more complex and connected world must be having a profound effect on the wiring of our brains.

Science-fiction writers have often envisioned humans of the far future as having big brains and stick-like, atrophied legs. But this is to ignore the lesson of fossil history.

Cro-Magnons, our ancestors in Europe, had bodies and brains between 5 and 10 per cent bigger than ours. The reason for this might have been that a bigger body was a stronger body, able to protect itself and ensure survival. And, every second of every day Cro-Magnons had to worry about survival, whereas today many of us live in a more benign world where someone else does the hunting, someone else supplies the food. Tellingly, domesticated animals invariably have smaller brains than their wild cousins. 'Through culture, humans effectively domesticated themselves,' said paleoanthropologist Louis Leakey.[9]

Whatever the reason for the rapid shrinking of people after Cro-Magnons, the trend is clear. Contrary to expectations, the humans of the future will probably have brains that are not bigger than ours but significantly smaller. Of course, whether we actually have a future depends on our solving a multitude of global problems, many of our own making. It is here that we confront the sobering lesson of our past. For as the American biologist Edward O. Wilson said, 'We have created a Star Wars civilisation but we have Stone Age emotions.'[10]

PART TWO: Putting matter to work

7:

A LONG HISTORY OF GENETIC ENGINEERING

Civilisation

The first human who hurled an insult instead of a stone was the founder of civilisation.

SIGMUND FREUD

Culture is roughly everything we do and monkeys don't.

FITZROY SOMERSET, 4TH BARON RAGLAN

The human story did not of course finish with the ice ages. The ice spread down from the poles; the ice returned whence it came. Over and over again, ebbing and flowing like a mile-deep white tide. But then, after the ice had retreated and the sea level had risen for the umpteenth time, something changed in the world. The change was so profound that the train of events it set in motion could conceivably prevent the return of the ice. Not for a while. But for ever.

The development was, of course, farming.

For thousands of generations, humans had no choice but to eat what produce nature laid out on its table. As hunters, they had followed the great herds of game. As gatherers, they had picked fruit and berries from bushes and trees. But, around 8500 BC, in the south-west corner of Asia, there appeared something entirely new under the sun: a fresh and innovative way of living.

The Fertile Crescent is a band of biological abundance whose epicentre is the land between the Tigris and Euphrates rivers in present-day Iraq, Syria and Turkey. Among the plants that thrived there were wild grasses with big edible seeds. No doubt people had grazed on these cereals for as long as they had grazed on other berries and fruit. In fact, since the plants often grew in large stands, people might have begun to rely on them for a significant part of their diet.

Around 12,000 BC, the world was still in the grip of intense cold, but the ice age was beginning to falter, and its last millennium was punctuated by spells when the climate warmed just a little. During one of these spells, according to British archaeologist Chris Scarre, people began to do something they had never done before. The details are sketchy. Perhaps they uprooted cereals and replanted them in moist rich soil close to one of the big rivers. Or maybe they simply weeded out competing plants in order to leave a field that was exclusively covered in cereals.

In the beginning, the cereals selected for special treatment were indistinguishable from their wild versions. But something subtle that people did began to change everything. They planted only the wild cereals with the biggest seeds and the easiest-to-remove husks. And, when they harvested their cereals, they replanted only the cereals that had borne the biggest seeds and whose husks popped open most readily. Gradually, because of these actions, the seeds grew bigger and easier to harvest season by season.

Inadvertently, without the slightest idea of what they were doing to the DNA of their crops, the farmers of the Fertile Crescent had become the first genetic engineers, exponents of evolution by human, or artificial, selection. Natural selection was not completely cut out of the equation. The plants that thrived were precisely those that could best tolerate the artificial environments created by humans – plants that grew fast even when exposed to the baking sun in unshaded fields or which grew strong when packed close together in rows. But, whether it was natural selection that was operating or human selection, *it was operating to a human agenda*. For the first time in history, people were directing the evolution of other species.

By 8500 BC, the cereals had diverged so significantly from their wild cousins that they were a different species. Wheat had become domesticated. And it was not alone. By 8500 BC, people had domesticated pea and olive too. And these were just the first of many domesticated crops that would follow over succeeding millennia. Modern agriculture had begun and nothing in the world of humans, *nothing in the world*, would be the same again.

An obvious question is why did farming start in the present interglacial and not in the previous one, between 130,000 and 115,000 years ago? A possible answer is that humans at the time had not yet evolved the intellectual capacities and imagination of fully modern humans. The sudden flowering of art across the world around 50,000 BC is often taken as evidence of a profound change in the wiring of human minds.

Communities, chiefs and cuneiform

The changes brought about by the shift to growing food were profound. For the first time in history, people were able to live their whole lives in one location. Instead of brief overnight halts around campfires, they were able to stop for good in permanent settlements. That is not to say that that there had been no settlements before. People might have been able to stay put if, for instance, they were close to an abundant source of food such as a fish-filled lake. Farming, however, made it possible for a lifestyle that had been a rare exception to become widespread.

But the key change was not the creation of settlements but the creation of a food surplus. Farming can typically feed between 10 and 100 times as many people per square kilometre as hunting and gathering. The surplus of food meant more people could be supported, which meant the population could grow. Ironically, this growth inevitably meant that the food was spread ever more thinly, so, in the long term, people might actually have been more badly nourished than their hunter-gatherer ancestors. But, by this time, there were too many people per square kilometre to be supported by the old lifestyle. The point of no return had been passed and there was no going back.

The surplus of food meant not only that more people could be fed; it meant that not everyone needed to be engaged in obtaining food. For the first time it was possible to support non-productive people such as craftsmen who made bricks or pottery or jewellery.[1] Of course, craftsmen could not exist without a market for their goods. And here the new developments fed off each other. The sedentary lifestyles permitted people to clutter their homes with belongings. By contrast, hunter-gatherers, continually on the move, were limited to what they could carry – a baby or a handful of spears.

But the food surplus, in addition to supporting craftsmen, could also support soldiers whose job was to defend a settlement and its fields. And not only soldiers but a chief. Hunter-gatherer societies tend to be egalitarian.[2] However, settlements of many people are complex. And, just as the myriad functions of a cell need to be orchestrated by a central nucleus, the myriad functions of a village need to be orchestrated by some kind of central government. The control of people by such a ruling elite, with soldiers at its disposal, provided both dangers and opportunities. Inevitably, it would lead to conflicts over land and resources. But it would also spawn cities and, eventually, empires of hundreds of thousands, even millions, of souls.

Such vast accumulations of people were an extraordinary and unprecedented thing. Our great-ape cousins live in small bands and react violently to outsiders. Initially, humans might have shared their xenophobia. But, as communities grew in size, people had to overcome their knee-jerk instinct to lash out at others. Benjamin Franklin spelled out the recipe for living in large communities: 'Be civil to all; sociable to many; familiar with few.' The key importance of this was recognised by Sigmund Freud. 'It is impossible to overlook the extent to which civilisation is built up upon a renunciation of instinct,' he wrote.[3]

Natural selection was probably operating here too. Those settlements where people were best at living in close proximity to each other without running amok had the lowest death rates and so grew faster than others. Their numerical domination meant that, over time, people became ever more passive and tolerant of each other. This is the view of Canadian-born psychologist Steven Pinker, who argues that, despite millions dying in the world wars of the twentieth century, the human race has shown a marked trend towards becoming less violent and warlike.[4] 'Civilisation is just a slow process of learning to be kind,' wrote Tennessee Williams.

People might have actually gone so far along the road to getting on with each other that they now choose to flock together in cities where they can interact with many others. 'People *like* being with other people,' says Scarre. Charles Dickens, who found himself living at a unique time when nineteenth-century London was making the transition to a mega-city of many millions, was one of the first to notice the new opportunities for interactions between large numbers of people. It might explain why in his novels chance encounters play such a big role.

The growth of societies and the growth of their complexity eventually triggered another profound development in human history: the invention of writing. This did not happen overnight. At first, the wedge-shaped marks made on clay tablets by the Sumerians around 3400 BC recorded only dull commercial transactions. But basic cuneiform was superseded by written languages that could express non-utilitarian things such as thoughts and feelings. Just as the cooking pot acted as an external stomach, the written word acted as an external memory. Whereas only a small amount of knowledge could be transmitted between people verbally, an enormous amount could be now transmitted in written form. By means of writing, the human race acquired a collective brain. Its full awakening would require widespread literacy, which would take many millennia. But, even in the earliest days, the writing was on the wall (pun intended).

Animal magic

So far, I have not mentioned animals. But, of course, the first genetic engineers did not simply manipulate plants. They captured and tamed large mammals and bred them for passivity or meat content. In the process, animals too diverged from their wild cousins and became domesticated. Not only did they provide food but they provided the motive power to pull ploughs to break up the hard soil and to pull goods on wheeled carts. The first animals to be domesticated, again in the Fertile Crescent, were sheep and goats around 8000 BC. In China, pigs and silkworms were both domesticated by 7500 BC, along with rice and millet.

The importance of animals to humans cannot be overestimated. Everywhere in the world where there are people animals live with them, not just as walking larders but also as pets. The American anthropologist Pat Shipman argues that our relationship with animals is central to understanding the extra-ordinary success of humans.[5] She points out that, from the moment humans first became artists around 50,000 BC, they rarely depicted themselves or their environment. Instead, almost exclusively, they painted prey animals. Their stunningly vivid representations, perfect in every anatomical detail, reveal what fantastically keen observers of animal behaviour they were. And it was this intense studying of animals, Shipman argues, that enabled humans to out-think their prey. It is the key, she believes, to understanding how a puny and insignificant ape, outrun and outgunned by big predators on the African savannah, managed to gain such control over its world.

Controversially, Shipman even believes that language might have arisen in order to exchange knowledge about animals. She believes this explains why humans domesticated an animal such as the dog, which is rarely eaten and which actually competes for the very same food resources with humans. Dogs were domesticated at least 17,000 years ago and possibly as early as 32,000 years ago.[6] The fact we keep pets, Shipman believes, reveals a profound important truth: without animals, humans are not humans.

Together, the domestication of plants and animals can be categorised as the invention of food production. This was the post-ice-age revolution that transformed the human race. Every-thing that has happened since has been a consequence of it. The birth of a sedentary lifestyle, first in villages, then in cities. The birth of specialised jobs. The creation of ruling elites. The birth of writing. The invention of war and creation of empires.

In 8500 BC, a ball was set rolling that was unstoppable and that to this day is continuing to gather momentum.

Modern milestones

There have been so many milestones along the road to the world of the twenty-first century that it is hard to know what to mention and what to leave out. But some of the most important developments occurred from the late fifteenth century onwards. Very significant was the advent of ships that could cross oceans, connecting Europe to the Americas, Africa, Asia, and finally Australia. This marked the beginning of a truly global civilisation. Later, towards the end of the eighteenth century, world trade was boosted enormously by mechanised factories that could mass-produce goods. This industrial revolution was powered first by water, then by coal, an energy source around 150 times more potent than human muscle power. Having access to such energy sources is one of the reasons that only a few per cent of the population are able to provide food for the overwhelming majority.

A development just before the industrial revolution, however, was also hugely significant. The rise of science, in the seventeenth century, illustrates how the collisions between ideas or technologies can spawn new ideas and new technologies. For thousands of years there had been craftsmen. By getting their hands dirty, they discovered how to make harder swords or better clay pots. But they worked by trial and error and never created a *theory* of what they were doing to guide them in finding ways to do it better. Besides the craftsmen, there were natural philosophers. Beginning with the Greeks more than two and a half millennia ago, they theorised loftily about the world but did not get their hands dirty in order to test their theories. What changed in the seventeenth century was that these two separate rivers of human expertise flowed together and became an unstoppable flood: science.

Isaac Newton epitomised the merger of the craft and philosophy traditions. He looked at the world, theorised about why it behaved the way it did, then got his hands dirty by carrying out experiments to test his theories against reality. The idea that by observing the world systematically it was possible to gain new knowledge was extraordinarily productive. It has led us to aeroplanes and antibiotics, cars and computers, neutron stars and nuclear reactors.

Why here and not there?

Science, the industrial revolution, ocean-going ships, and so many other things, were born in Europe. Together, such technologies ensured that it was Europeans who colonised the Americas and Australia and ultimately spawned our modern global society. This prompts a question. It was asked by the American geographer Jared Diamond in his book *Guns, Germs and Steel*: 'Why did human development proceed at such different rates on different continents for the last 13,000 years?'[7]

Diamond believes it was not because of any intrinsic differences in human beings – people in the Americas and Australia were not more stupid or more lazy than their European counterparts – but because of differences in their circumstances. Together, Europe and Asia – the source of so many of Europe's innovations – had a far bigger and more diverse population than either the Americas or Australia. Not only were there more societies to interact with each other but there were more individuals within each society interacting with each other. The overall effect was to boost the exchange of ideas and accelerate the rate at which people invented new things.

This, of course, prompts yet another question: why did Europe and Asia have a bigger population than the Americas and Australia? The answer, according to Diamond, was because they had a big head start in food production. This, in turn, was because the Fertile Crescent had a wider range of habitats, from deserts to rich soils to snowy mountaintops, which created a super-abundance of different plant species. At least a dozen of these were candidates for domestication compared with only a couple in, for instance, the Americas. These domesticated species were carried by farmers as they migrated from south-west Asia to Europe. Ultimately, the dominance of European culture comes down not to any intrinsic superiority of Europeans but to a mere accident of birth.

Even with the domestication of animals, Europe and Asia had significant advantages over the Americas and Australia. The ultimate reason for this actually pre-dates even the birth of modern agriculture in the Fertile Crescent. It has to do with the date at which humans arrived on different continents. People flooded into Europe and Asia very early in human history. Consequently, they carried unsophisticated stone tools and were not particularly effective hunters. Animals were able to live alongside them for a long while and learn to be fearful of them. This ensured the survival of the big mammals that one day would be candidates for domestication.

The Americas and Australia, on the other hand, were reached relatively recently by humans – Australia in about 50,000 BC and the Americas only in about 14,000 BC. Consequently, the first colonists were fully modern. Far from being unsophisticated hunters, they were lethal killers. And, sure enough, the arrival of humans in both Australia and the Americas appears to have coincided with the extinction of pretty much all of their giant mammals with the exceptions of the bison in North America and the llamas and alpacas in the Andes. The result of this was that, thousands of years later, when animals were first being domesticated in the Fertile Crescent, there were few suitable candidates for domestication in Australia and the Americas.

The huge disadvantages people in the Americas and Australia faced in domesticating crops and domesticating animals is why their populations did not grow big enough to permit the feverish level of interaction necessary to create novel inventions. And this explains why it was the Spanish who sailed across the ocean to South America and not the Aztecs and Incas who sailed in the opposite direction to Europe. Furthermore, the empires of South America did not have horses, steel armour or guns, which meant that bands of a few dozen mounted Spaniards were able to rout native armies numbered in thousands.

In North America, the Native Americans who met the first Europeans were at an even bigger disadvantage than their cousins in the south of the continent. They possessed only weapons of stone and wood and no animals that could be ridden.

In the human catastrophe that unfolded, something like 95 per cent of the native people of the Americas were wiped out. Although many succumbed to guns, it was not actually superior technology that killed most of the people in the Americas, not to mention Australia. It was diseases such as smallpox and measles brought by the Europeans. This poses yet another puzzle. 'It's striking', says Diamond, 'that Native Americans evolved no devastating epidemic diseases to give to Europeans, in return for the many devastating epidemic diseases that Indians received from the Old World.'

Yet again, the explanation has to do with the head start Europe and Asia enjoyed in food production. Many human diseases originate in common domestic animals such as pigs and chickens. Over thousands of years, the farming of many more animals, and many more types of animals, had created many more animal diseases. Occasionally, these spread to people and became human diseases. Measles and tuberculosis, for instance, evolved from diseases of cattle, flu from a disease of pigs, and smallpox possibly from a disease of camels. The Americas, by contrast, had very few native domesticated animal species from which humans could catch diseases.

The animal diseases that adapted themselves to humans in Europe and Asia ravaged the large population. Countless millions died but the survivors were left with immunity – immunity that, crucially, the conquered peoples of the Americas and Australia simply did not have. 'Civilisation is what makes you sick,' observed artist Paul Gauguin.

The differences between human societies on different continents, according to Diamond, are not down to any biological differences between people but down to differences in continental environments. Mark Twain recognised this even in the nineteenth century. 'There are many humorous things in the world,' he wrote, 'among them the white man's notion that he is less savage than the other savages.'[8]

Footprints in the dust

Looking back over the past 13,000 years since the end of the last ice age, it is clear what the driving force of most human innovation has been: Interaction. Interaction. Interaction. The settlement of people, first in villages, then cities, boosted the opportunities for people to exchange ideas and ramped up the rate of technological advance. Today, we have a global civilisation with more than 7 billion people and the confluence of computers and telecommunications has spawned the internet, which has exponentially boosted the number of interactions between people. In 2012, the number of text messages sent a year was estimated to be a staggering 8.7 *trillion*.[9]

But things are not looking good for the human race. Not only do we have the ability to destroy our global civilisation in a single day with nuclear weapons but our sheer numbers are putting the global environment under creaking strain. The climate is changing, the seas are losing their productivity and the species we share the planet with are suffering a major extinction event. Not since the advent of cyanobacteria, which poisoned the planet with oxygen, has a single species had such a devastating effect on the Earth. All we can hope is that the unprecedented level of interactions between human beings will throw up the solutions we need to head off a catastrophe.

Our extinction now would be a terrible shame because we have come so far and have achieved so much. Perhaps the most extraordinary development occurred on 20 July 1969 when a human being for the first time set foot on another world. In the annals of life on Earth, Neil Armstrong's 'one small step for [a] man, one giant leap for mankind' was the most significant development since the first amphibian crawled out of the ocean onto dry land 350 million years ago. Who would have guessed, when *australopithecines* left footprints in the Laetoli dust that, 3.6 million years later, their descendants would leave footprints in the dust of the Sea of Crises?

But let us not get carried away. Let us remember why we are here: because our farming ancestors learned the subtle art of genetic engineering. As an anonymous writer observed, 'Man – despite his artistic pretensions, his sophistication, and his many accomplishments – owes his existence to a six-inch layer of topsoil and the fact that it rains.'

8:

THANK GOODNESS OPPOSITES ATTRACT

Electricity

We believe that electricity exists because the electric company keeps sending us bills for it, but we cannot figure out how it travels inside wires.

DAVE BARRY

Is it a fact — or have I dreamt it — that, by means of electricity, the world of matter has become a great nerve, vibrating thousands of miles in a breathless point of time?

NATHANIEL HAWTHORNE, *The House of the Seven Gables* (1851)

A thin metal wire comes into your home and something *invisible* travels down it. The *something* not only has the *oomph* to spin the tumbler of a washing machine but to light every room in your home – and in winter even heat your home as well. Not only your home but millions of other homes. *Billions* even. Everyone knows that electricity powers the planet. But how in the world does it do it?

Here's an explanation. It requires a little background. Imagine there is a force that behaves like gravity but differs from gravity in two key respects.[1] First, instead of separate chunks of matter always attracting each other in the way that the Sun and Earth do, there are *two* types of matter that experience the force differently. Call them Type 1 and Type 2, or A and B, or positive and negative. It does not matter. The key thing is that *unlikes* attract with the force while *likes* repel with the same force.

A bunch of positives therefore repel and flee from each other in all directions and a bunch of negatives does exactly the same. However, with an evenly mixed bunch of positives and negatives, something quite different happens. The opposite pieces pull each other together and the like pieces drive each other apart. But, because the opposing forces are equal and opposite, there is a perfect balance.

It follows that if there are two bodies, each of which is an equal mixture of negatives and positives, they will neither attract nor repel each other.

A force that behaves like this does indeed exist. It is called the *electric force*. And ordinary matter – the stuff of which you and me and the world around us is made – turns out to be an even mixture of positively *charged* protons and negatively *charged* electrons. (Protons are confined to the core, or nucleus, of each atom, whereas electrons orbit the nucleus.) The balance of attraction and repulsion is so perfect that, when you stand near someone else, neither of you feels the slightest force. In fact, in everyday life, there is very little hint that the electric force exists at all.

A force that can be both attractive and repulsive but which, in normal circumstances, is cancelled out perfectly probably seems dull and unremarkable. But, remember, the electric force differs from gravity in not one but *two* respects. While the first difference ensures that the force is pretty much always nullified, the second difference is at the very root of the force's extraordinary ability to power the modern world. The electric force is *stronger* than gravity. But not by a factor of 10. Or of 100. Or even a million. No, the electric force is stronger than the force of gravity by a factor of *10,000 billion billion billion billion*.[2]

To get some idea of what this enormous number means, imagine a mosquito buzzing in a jar. Say, by some wizardry, it is possible to remove all the negative electrons from the atoms of the mosquito so that all that are left behind are the positive atomic nuclei.[3] These will, of course, repel each other. The mosquito will explode. The question is: with how much energy will the mosquito explode?

(a) The energy of a sparkler?

(b) The energy of a stick of dynamite?

(c) The energy of a 1-megatonne H-bomb?

(d) The energy of a global mass extinction?

Perhaps you think the answer is (b) a stick of dynamite, or maybe (c) a 1 megatonne H-bomb? If you think (c), you are at least on the right track. A hydrogen bomb is a useful comparison. But not a *single* hydrogen bomb. *A million billion 1-megatonne H-bombs.* The mosquito will explode with an energy equivalent to the city-sized asteroid that slammed into the Earth 65 million years ago and wiped out the dinosaurs. The answer is (d). The mosquito will explode with the energy of a *global mass extinction*. Were it not for the fact that the mind-bogglingly huge electric force – 10,000 billion billion billion billion times stronger than gravity – is invariably cancelled out, each and every mosquito on Earth would be a potential world-destroyer. Thank goodness that in physics, as in life, opposites attract.

Now perhaps it is possible to appreciate the potential of the electric force for energising the world.

Removing all the electrons from a mosquito – if it were possible – would create a dramatic charge imbalance and unleash a truly extraordinary amount of electric energy.[4] It follows that creating even a modest charge imbalance might unleash a significant amount of electric energy. This is what happens in a thunderstorm. Here, a charge imbalance builds up between a cloud and the ground (or, more commonly, between one cloud and another). Specifically, the underside of a cloud builds up a negative charge at the expense of the ground, which becomes positively charged. Eventually, the electric force between the cloud and the ground becomes so immensely strong that it is able to rip the outer electrons from the atoms in the air between.

This breakdown of the air sends an avalanche of electrons – typically, 100 billion billion of them – surging down to the ground to cancel out the charge imbalance. In short, it creates a lightning bolt.

A flow of electrons is called an electric current.[5] Typically, in the case of a lightning bolt, the current is about 10,000 amps, though it can be as high as a few hundred thousand amps (by comparison, many household electrical appliances use less than 10 amps). For just a tenth of a second or so, the current surges down a channel the width of a pencil.[6] The electrons that compose it slam into air atoms, like a myriad tiny ball bearings, transferring energy to their still-bound electrons. The air atoms gain so much energy that the temperature can soar to about 50,000 °C – almost 10 times hotter than the surface of the Sun. It is the supersonic expansion of this blisteringly hot air on either side of the lightning channel that creates the clap of thunder. And it is the atomic electrons, shedding their excess energy as photons, that *light up* the lightning bolt.

Lightning demonstrates some key properties of electricity. One is that, if a charge imbalance is created, the electric force is presented with an opportunity to unleash a large amount of energy.[7] Another is that electrical energy can be transferred *across a distance* by an electric current. In a lightning bolt, the distance is typically a couple of kilometres. However, the longest recorded streak of lightning, observed near Dallas, Texas, was almost 200 kilometres long. The ability to transfer energy across a distance by an electric current has been of huge significance in creating the modern technological world.

Lastly, of course, lightning demonstrates that, by means of an electric current, it is possible to change electrical energy into other forms of energy – specifically, heat and light. The 'killer app', responsible for kick-starting the electrical revolution of the late nineteenth century, was in fact the light bulb. The electrical pioneers were not thinking about putting electricity into the home or even electrical appliances in the home. They were thinking about putting *light in the home*. 'The light bulb is what wired the world,' says Jeff Bezos, founder of Amazon.com.

Just as a current in lightning transfers energy to the air – heating and lighting it up – a current in a light bulb transfers energy to a filament – heating and lighting it up. The clever bit – perfected, though not invented, by Thomas Edison – is putting a filament in an oxygen-free glass bulb so that it glows *without burning away*.[8]

But although lightning demonstrates some key properties of electricity, building up a huge charge imbalance and waiting for the air to break down catastrophically is hardly a practical way to generate an electric current. Fortunately, there is a more convenient and controlled way. To understand it, however, it is first necessary to appreciate how exactly the electric force of an electric charge reaches out across space and influences other charges.

The electric force field

If you rub a balloon against a nylon sweater, loose electrons get transferred from one to the other. It does not matter which way they go – and, in fact, it is not entirely clear. The point is that both the balloon and sweater become electrically charged.[9] If you now bring the charged-up balloon close to a small scrap of paper, the scrap will leap through the air, yanked by the electric force, and glue itself to the balloon.[10] Somehow, the electric force of the balloon has reached out through the air and grabbed the scrap of paper.

Physicists say that extending out through space from an electric charge is an invisible electric *force field*, rather like a *Star Trek* tractor beam. When the paper finds itself in the field, it experiences a force towards the charge.[11]

The field of force around a charged balloon is feeble but between a storm cloud and another cloud or between a cloud and the ground it can be enormous. And it is this field that eventually becomes so irresistibly strong that it tears electrons from the very atoms of the air, creating the electron avalanche of a lightning bolt. In fact, the field in a thunderstorm can be strong enough to be *felt*, prickling the skin and even making hair stand on end. Mind you, if you experience either of these sensations, throw yourself flat to the ground. A lightning strike is imminent and your name is written on it.

The magnetic force field

But there is more to the electric field than an invisible force field that extends outwards from an electric charge (pulling in unlike charges and pushing away like charges). This merely describes the force surrounding a *static* charge. If the charge is *moving* relative to a second charge, a new force puts in an appearance. The second charge, in addition to the electric force, experiences a *magnetic force*.

The magnetic force field is easier to imagine than an electric force field. After all, if you have a bar magnet and a nail and bring them together, you can actually *feel* the invisible tractor beam of the magnetic field of the magnet clamping onto the nail. In fact, it was seeing the needle of a magnetic compass respond to the Earth's magnetic field that blew the mind of Albert Einstein, aged four or five, switching him on to science and teaching him a lesson about nature that he never forgot: there is 'something behind things, something deeply hidden'.[12]

The fact that a magnetic field is caused by a *moving electric charge* helps explain the origin of the magnetic field of permanent magnets. Every material, including the flesh and blood of which you are made, consists of countless charged electrons, not only moving in orbit around the nuclei of atoms but actually behaving like tiny spinning tops themselves. This means every atom and every electron is like a tiny magnet. In most materials, all the countless mini-magnets are orientated randomly and so, overall, their magnetic fields cancel out. However, in some materials, this cancellation is not perfect. Such materials are permanent magnets.

The fact that a moving electric charge creates a magnetic field was first noticed in 1820 by Danish physicist Hans Christian Ørsted. He saw that the needle of a magnetic compass was deflected when he brought it close to a conducting wire carrying an electric current. A current, by definition, is electric charge in motion, and electric charge in motion obviously has a changing electric field. What Ørsted realised was that *a changing electric field creates a magnetic field*.

Bring two magnets together and feel the powerful force between them.[13] As Ørsted discovered, a current-carrying coil of wire, with its changing electric field, *is* a magnet. Bring it together with a permanent magnet and there will be a force between the two, just as there would be between two permanent magnets. Arrange the coil and the magnet in the right way – and this takes some ingenuity — and the force will cause the coil of wire to *spin*. *Voilà*. You have created an electric motor.

The reason a magnetic field can spin something is that it has what physicists call curl. Whereas an electric field extends radially outwards from a charge, a magnetic field — for instance, one created by a current-carrying wire — swirls around like a miniature tornado of force.

With the aid of an electric motor, it is possible to do a lot more with an electric current than merely create heat and light. It is possible to *move* things. In the motor of an electrical appliance, the changing electric field of an electric current generates a magnetic field of the force that propels a spindle. It is possible to drive everything from washing machines to automatic doors to electric cars and trains. And this is all down to the simple fact that a *changing electric field creates a magnetic field*.

Of course, it goes without saying that the prerequisite for running an electric motor is an electric current. Nature can create one — fleetingly and chaotically — in a lightning bolt. But how is it possible to create an electric current in a practical and controlled way? The answer is by exploiting another property of electric and magnetic fields. It was first noticed in 1831 by English physicist Michael Faraday. Faraday was the father of our electrical power system. 'Even if I could be Shakespeare I think that I should still choose to be Faraday,'[14] wrote Aldous Huxley, author of *Brave New World*. And, famously, when asked by William Gladstone, Chancellor of the British Exchequer, 'What is the practical use of electricity?', Faraday replied, 'Why, sir, there is every probability that you will soon be able to tax it.'

Faraday noticed that, when he moved a magnet in the vicinity of a coil of conducting wire, an electric current flowed fleetingly in the coil.[15] A current is a flow of charges, and charges are propelled by an electric field. What Faraday had noticed was that *a changing magnetic field creates an electric field*.

There is a pleasing symmetry between electric and magnetic fields. Not only does a changing electric field create a magnetic field but a changing magnetic field creates an electric field. Spin a coil of wire in the magnetic field of a permanent magnet. Arrange the coil and magnet in the right way – and this also takes some ingenuity – and an electric field will be created in the coil, which will drive a current. *Voilà*. You have created an electric generator.

At a power station, a coil of wire is spun in the force field of a magnet. The thing doing the spinning could be wind or water, or steam from water heated by coal or oil or nuclear power.[16] The key thing is that the spinning coil cuts through the magnetic field. In other words, the magnetic field through the coil *changes*. And the changing magnetic field creates an electric field, which pushes electrons around the coil. It creates a current. The same current that surges out of the power station down a cable and powers your home.

In practice, the current goes in and out of each home *in the same cable*. The current arrives from the power station in the live wire and returns to the power station in the neutral wire, thus completing the circuit.[17] There is a twist, however. Although the statement above is basically true, power stations do not generate current that flows in one direction only. Instead of direct current, or DC, they generate alternating current, or AC, which sloshes back and forth, rapidly changing its direction. The reason for this is to overcome a major problem with the long-distance transmission of electricity.

Alternating current

Think of an electric current flowing from a power station like a stream flowing down a hillside to a valley bottom. If you were to intercept the stream near the top of the hill, you would be able to exploit the long drop of the water to the valley bottom to power a piece of machinery such as a water wheel. However, if you were to intercept the stream close to the valley bottom, there would be very little drop left that could be exploited. And this is also the problem with an electric current flowing from a power station. The further away a home, the less energy can be extracted from the flow of electrons. While those close to the power station might be able to light their homes with a multitude of bright light bulbs, those far away will have to make do with the faintest of glimmers from a single light bulb.

The amount of energy a current can deliver is characterised by its voltage, which is analogous to the height of that stream above the valley bottom. In the UK, domestic appliances work on 240 volts (110 volts in the US). This would seem to imply that a power station would have to generate electricity at near 240 volts. But, of course, if it did, those living far away might have to make do with 100 volts, or 10, or even a measly 1 volt. In New York in the 1880s, the only way Edison could overcome this Achilles heel of electric power transmission was to build a power station about every 2.5 kilometres. Although this is just about doable, it is clearly an unworkable solution for distributing electrical power nationally.

The solution to the transmission problem, pioneered by Nikola Tesla and others, is to generate the electrical power not at 240 volts but at a voltage about *a thousand* times greater. In the UK's

National Grid, electricity is transmitted over long distances at 110,000 volts or higher.[18] This means that, if the electrons lose energy travelling down the wire from a power station – and they inevitably do, banging into atoms and losing energy as heat – the voltage drop is hardly noticeable compared with 110,000 volts. Consequently, it is possible to transmit electricity over huge distances, and power stations do not have to be near homes. The problem is that 110,000 volts is too enormous for household appliances. Somewhere between a power station and people's homes, the voltage must reduced, or stepped down. This is not possible with a direct current. But it is possible with an *alternating current*.

An alternating current switches direction rapidly, commonly between 50 and 60 times a second. Think of the electrons in a wire as sloshing back and forth like waves on a sea shore. And, since an alternating current is ultimately made of countless *moving* electrons exactly like a direct current, it can carry energy as efficiently as its uni-directional cousin.[19] Furthermore, it is possible to design both a generator to *create* alternating current and a motor to *run on* alternating current. There remains only the problem of stepping down an AC current from 110,000 volts to the 240 volts required domestically. But this can be done with a transformer.

In a transformer, a current changing in one coil of wire – that is, an alternating current – causes a changing magnetic field in a second coil. This, in turn, creates a changing electric field, which drives a changing current in the second coil. If the second coil has fewer *turns* than the first, then the voltage *goes down*.

So there you have it: our electrical power system in a nutshell.

Electricity, magnetism and light

But electricity and magnetism have some more tricks up their sleeve. Recall that, if an electric charge is moving relative to you, you see not only an electric field *but a magnetic field*. However, if you travel alongside the charge, so that it is not moving relative to you, you see no magnetic field. And this is not all. If a magnet is moving relative to you, you see not only a magnetic field *but an electric field*. But, if you travel alongside the magnet, you see only a magnetic field.

How is it possible that, from one perspective, there is a magnetic field and, from another, no magnetic field? How is it possible that, from one perspective, there is an electric field and, from another, no electric field? There is only one way it can be possible: if magnetic fields and electric fields are *not fundamental things at all.*

As Einstein realised in 1905, an electric field and a magnetic field, just like space and time, are simply different facets of the same thing – an electromagnetic field. How much of each facet you see depends on your speed relative to the source of the electromagnetic field. This is why what one person sees as an electric field someone else sees as an electric field *and a magnetic field.* This is why what one person sees as a magnetic field someone else sees as a magnetic field *and an electric field.* No wonder there is a pleasing symmetry between the behaviour of electric and magnetic fields. How can there not be? They are essentially the *same* thing.

But there is yet more. In 1863, the Scottish physicist James Clerk Maxwell, in a scientific tour de force, distilled all known electrical and magnetic phenomena into a single neat set of equations.[20] Studying those equations, he noticed something remarkable. It appeared possible for a ripple to propagate through the electric and magnetic fields just like a wave on a lake. Not only was the ripple self-sustaining but it travelled at a very particular speed: *the speed of light*.

Maxwell had discovered a surprising connection between electricity and magnetism *and light*. Light, it turns out, is a wave of electromagnetism – an electromagnetic wave.[21] And there is more. Maxwell's equations reveal that it is possible to have electromagnetic waves that oscillate both more rapidly and more sluggishly than visible light. In 1888, their existence was proved by the German physicist Heinrich Hertz. With the aid of a spark, he transmitted an electromagnetic wave. The invisible-to-the-naked-eye radio wave crossed his laboratory and induced a measurable current in a coil of wire. It was an epochal, world-changing moment. All radio and television communications around the world began with that one triumphant demonstration. Our connected modern world was born that day. 'From a long view of the history of mankind, seen from, say ten thousand years from now, there can be little doubt that the most significant event of the nineteenth century will be judged as Maxwell's discovery of the laws of electrodynamics,' said American Nobel Prizewinner Richard Feynman.[22]

Electricity and the realm of the atom

Electricity opened up technological possibilities that were unimaginable to earlier generations. Not only was it possible to transmit a signal around the globe so that one person could talk to another without any material connection existing between them but it was possible to run huge extended power systems. In the evocative words of Feynman, 'Ten thousand engines in ten thousand places running the machines of industries and homes – all turning because of the knowledge of electromagnetism.'[23]

But, hand in hand with the harnessing of electricity came a dawning realisation that electricity is of central importance in nature. We live in an electrical world. Nobody had realised this before because, in pretty much all everyday circumstances, the enormous electric forces are perfectly balanced and so nullified. This is not the case, however, in the realm of the atom, the building block of all matter. There charge imbalances are ubiquitous.

As the joke goes . . .

Two atoms are walking down the street when they collide. One says to the other, 'Are you all right?'

'No, I lost an electron.'

'Are you sure?'

'Yeah, I'm positive.'

In a piece of matter with only a few atoms, there will usually not be an equal number of positive and negative charges. And, even if there are, there might still be large electrical forces. This is because the negative charge of one piece of matter might be closer to the positive charge of another piece of matter than its negative charge. Since the electrical force weakens with distance, attraction will win out over repulsion. Thus it is possible for two small pieces of matter to attract each other fiercely even if neither has a net charge.

Atoms, it turns out, are totally dominated by the immensely strong electric force. The glue that holds them together, and sticks them to other atoms to make molecules, is the electrical force. All chemistry, which involves the rearrangement of electrons in atoms, is electrical. The attraction of the electrical force not only holds the atoms in the molecules of your body together but the repulsion between the electrons on the outside of those molecules keeps you rigid, preventing the Earth's gravity from crushing you flat. And your cells have learned how to tap the energy of the electric force. Electrons from food create electric fields across cell walls that drive the creation of power-pack molecules such as adenosine triphosphate, or ATP. And electrons help to store and carry our very thoughts.[24]

Biology runs on electricity. We are electrical beings. We are as animated by the electrical force as much as a battery-powered toy is animated by the electrical force. Which explains why electricity is not only miraculous but *dangerous*. 'My nephew tried to stick a penny into a plug,' said American comedian Tim Allen. 'Whoever said a penny doesn't go far didn't see him shoot across that floor.'

9:

PROGRAMMABLE MATTER

Computers

One day ladies will take their computers for walks
in the park and tell each other, 'My little computer
said such a funny thing this morning.'
ALAN TURING

I think the world market for computers is maybe
. . . five.
THOMAS J. WATSON, Chairman of IBM, 1943

'Computers are useless,' said Pablo Picasso. 'They can only give you answers.' But what answers they give! In the past half a century, those answers have dramatically changed our world.[1] The computer is unlike any other human invention. A washing machine is a washing machine is a washing machine. It is impossible to change it into a vacuum cleaner or a toaster or a nuclear reactor. But a computer can be a word processor or an interactive video game or a smart phone. And the list goes on and on and on. The computer's unique selling point is that it can *simulate any other machine*. Although we have yet to build computers that can fabricate stuff quite as flexibly as human beings, it is merely a matter of time.[2]

Fundamentally, a computer is just a shuffler of symbols. A bunch of symbols goes in – perhaps the altitude, ground speed, and so on, of an aeroplane; and another bunch of symbols comes out – for instance, the amount of jet fuel to burn, the necessary changes to be made in the angle of ailerons, and so on. The thing that changes the input symbols into the output symbols is a program, a set of instructions that is stored internally and, crucially, is *infinitely rewritable*. The reason a computer can simulate any other machine is that it is *programmable*. It is the extraordinary versatility of the computer program that is at the root of the unprecedented, world-conquering power of the computer.

The first person to imagine an abstract machine that shuffles symbols on the basis of a stored program was the English mathematician Alan Turing, famous for his role in breaking the German 'Enigma' and 'Fish' codes, which arguably shortened the Second World War by several years.[3]

Turing's symbol shuffler, devised in the 1930s, is unrecognisable as a computer. Its program is stored on a one-dimensional tape in binary – as a series of 0s and 1s – because everything, including numbers and instructions, can ultimately be reduced to binary digits. Precisely how it works, with a read/write head changing the digits one at a time, is not important. The crucial thing is that Turing's machine can be fed a description of any other machine, encoded in binary, and then *simulate* that machine. Because of this unprecedented ability, Turing called it a Universal Machine. Today, it is referred to as a Universal Turing Machine.

Bizarrely, Turing devised his machine-of-the-mind not to show what a computer can do but *what it cannot do*. He was at heart a pure mathematician. And what interested him, even before the advent of nuts-and-bolts hardware, was the ultimate limits of computers.

Remarkably, Turing very quickly found a simple task that no computer, no matter how powerful, could ever do. It is called the halting problem and it is easily stated. Computer programs can sometimes get caught in endless loops, running around the same set of instructions for ever like a demented hamster in a wheel. The halting problem says: if a computer is given a computer program, can it tell, *ahead of actually running the program*, whether it will eventually halt – that is, that it will not get caught in an interminable loop?

Turing, by clever reasoning, showed that deciding whether a program eventually halts or goes on for ever is logically impossible and therefore beyond the capability of any conceivable computer. In the jargon, it is 'uncomputable'.[4]

Thankfully, the halting problem turns out not to be typical of the kind of problems we use computers to solve. Turing's limit on computers has, therefore, not held us back. And, despite their rather surprising birth in the abstract field of pure mathematics as machines of the imagination, computers have turned out to be immensely practical devices.

A vast numerical irrigation system

Computers, like the Universal Turing Machine, use binary. Binary was invented by Gottfried Leibniz, a seventeenth-century German mathematician who clashed bitterly with Isaac Newton over who had invented calculus. Binary is a way of representing numbers as strings of 0s and 1s. Usually, we use decimal, or base 10. The right-hand digit represents the 1s, the next digit the 10s, the next the 10 × 10s, and so on. So, for instance, 9217 means 7 + 1 × 10 + 2 × (10 × 10) + 9 × (10 × 10 × 10). In binary, or base 2, the right-hand digit represents the 1s, the next digit the 2s, the next the 2 × 2s, and so on. So, for instance, 1101 means 1 + 0 × 2 + 1 × (2 × 2) + 1 × (2 × 2 × 2), which in decimal is 13.

Binary can be used to represent not only numbers but also instructions. It is merely necessary to specify that *this* particular string of binary digits, or bits, means add; *this one* means multiply; *this one* execute these instructions and go back to the beginning and execute them again, and so on. And it is not only numbers and program instructions that can be represented in binary. Binary can encode *anything* – from the information content of an image of Saturn's rings sent back by the Cassini spacecraft to the information content of a human being (although this is somewhat beyond our current capabilities). This has led some physicists, drunk on the information revolution, to suggest that binary information is the fundamental bedrock of the Universe out of which physics emerges. 'It from bit,' as the American physicist John Wheeler memorably put it.[5]

Binary is particularly suitable for use in computers because representing os and 1s in hardware requires devices that can be set only to two distinct states. Take the storage of information. This can be done with a magnetic medium, tiny regions of which can be magnetised in one direction to represent a 0 and in the opposite direction to represent a 1. Think of an array of miniature compass needles. To manipulate the information, on the other hand, an electronic device that has two distinct states is necessary. Such a device is the transistor.

Imagine a garden hose through which water is flowing. The water comes from a source and ends up in a drain. Now imagine stepping on the middle of the hose. The flow of water chokes off. Essentially, this is all a transistor in a computer does.[6] Except of course it controls not a flow of water but a flow of electrons – an electrical current. And, instead of a foot, it has a gate. Applying a voltage to the gate controls the flow of electrons from the source to the drain as surely as stepping on a hose controls the flow of water.[7] When the current is switched on, it can represent a 1 and when it is off a 0. Simples.

A modern transistor (on a microchip) actually looks like a tiny T. The top crossbar of the T is the source/drain (the hose) and the upright of the T is the gate (the foot).

Now imagine two transistors connected together – that is, the source of one transistor is connected to the drain of another. This is just like having a hose beside which you and a friend are standing. If you step on the hose, no water will flow. If your friend steps on it, no water will flow either. If you both stand on the hose, once again no water will flow. Only if you do not stand on the hose *and* your friend does not stand on the hose will water flow. In the case of the transistor, electrons will flow only if there is a certain voltage on the first gate *and* the same voltage on the second gate.

It also possible to connect up transistors so that electrons will flow if there is a particular voltage on the first gate *or* on the second gate. Such *AND* and *OR* gates are just two possibilities among a host of logic gates that can be made from combinations of transistors. Just as atoms can be combined into molecules, and molecules into human beings, transistors can be combined into logic gates, and logic gates into things such as adders, which sum two binary numbers. And, by combining millions of such components, it is possible to make a computer. 'Computers are composed of nothing more than logic gates stretched out to the horizon in a vast numerical irrigation system,' said Stan Augarten, an American writer on the history of computing.[8]

Cities on chips

Transistors are made from one of the most common and mundane substances on the planet: sand. Or, rather, they are made from silicon, the second most abundant element in the Earth's crust and *one component* of the silicon dioxide of sand. Silicon is neither a conductor of electricity – through which electrons flow easily – nor an insulator – through which electrons cannot flow. Crucially, however, it is a semiconductor. Its electrical properties can be radically altered merely by doping it with a tiny number of atoms of another element.

Silicon can be doped with atoms such as phosphorus and arsenic, which bond with it to leave a single leftover electron that can be given up, or donated. This transforms it into a conductor of negative electrons, or an *n*-type material. But silicon can also be doped with atoms such as boron and gallium, which bond with silicon and leave room for one more electron. Bizarrely, the empty space where an electron *isn't* can move through the material exactly as if it is a positively charged electron. This transforms the silicon into a conductor of positive holes, or a *p*-type material.

A transistor is created simply by making a *pnp* or an *npn* sandwich – most commonly an *npn*. You do not need to know any more than this to grasp the basics of transistors (in fact, you already know more than you need).[9]

In the beginning, when transistors were first invented, they had to be linked together individually to make logic gates and computer components such as adders. But the computer revolution has been brought about by a technology that creates, or integrates, billions upon billions of transistors simultaneously on a single wafer, or chip, of silicon. The 'Very Large Scale Integration' of such integrated circuits is complex and expensive.[10] But, in a nutshell, it involves etching a pattern of transistors on a wafer of silicon, then, layer by layer, depositing doping atoms, microscopic wires, and so on.

To make a computer you need a computer. It is only with computer-aided design that it is possible to create a pattern of transistors as complex as a major city. Such a pattern is then made into a mask. Think of it as a photographic negative. By shining light through the mask onto a wafer of silicon, it is possible to create an image of the pattern of transistors. But a pattern of light and shadows is just that – a pattern of light and shadows. The trick is to turn it into something real. This can be done if the surface of the silicon wafer is coated with a special chemical that undergoes a chemical change when struck by light. Crucially, light makes the photoresistant material resistant to attack by acid.[11] So, when acid is applied to the silicon wafer in the next step of the process, the silicon is eaten away, or etched, everywhere except where the light falls. Hey presto, the image of the mask has been turned into concrete – or, rather, silicon – reality.

There are many other ingenious steps in the process, which might involve using many masks to create multiple layers, spraying the wafer with dopants and spraying on microscopic gold connecting wires, and so on. But, basically, this is the idea. The technique of photolithography quickly and elegantly impresses the pattern of a complex electric circuit onto the wafer. It creates a city on a chip.

Probably, most people think that microchips originate in the US or in Japan or South Korea. Surprisingly, they are born in Britain. The company behind the designs of the overwhelming majority of the chips in the world's electronic devices is based in Cambridge. ARM started out as Acorn Computers in 1985. While the big chip-makers like Intel in the US concentrated on making faster and more compact chips for desktop computers, or PCs, ARM struck out in a different direction completely. It put *entire computers* on a chip. This made possible the vast numbers of compact and mobile electronic devices from SatNavs to games consoles to mobile phones. It moved chips from dedicated and unwieldy computers *into the everyday world*.

Big bang computing

The limit on how small components can be made on a chip is determined by the kind of light that is shone through a mask. Chip-makers have made ever smaller components – packing in more and more transistors – by using light with a shorter wavelength, such as ultraviolet or X-rays, which can squeeze through smaller holes. They have even replaced light with beams of electrons since electrons have a shorter wavelength than light.[12] And chips have become ever more powerful.

In 1965, Gordon Moore, one of the founders of the American computer chip-maker Intel, pointed out that the computational power available at a particular price – or, equivalently, the number of transistors on a chip – appears to double roughly every eighteen months.[13] 'If the automobile had followed the same development cycle as the computer, a Rolls-Royce would today cost $100, get a million miles per gallon, and explode once a year, killing everyone inside,' observed Robert X. Cringely, technology columnist on *InfoWorld* magazine.[14]

People have been claiming that Moore's law is about to break down every decade since it was formulated. But so far everyone has been wrong.

Undoubtedly, however, Moore's law will break down one day. It is a sociological law – a law of human ingenuity. But even human ingenuity cannot do the impossible. There are physical limits set by the laws of nature, which are impossible to circumvent, that ultimately determine the limits of computers.

The speed of a computer – the number of logical operations it can perform a second – turns out to be limited by the total energy available.[15] Today's laptops are so slow because they use only the electrical energy in transistors. But this energy is totally dwarfed by the energy locked away in the *mass* of the computer, which provides nothing more than the scaffolding to keep a computer stable. The ultimate laptop would have all of its available energy in processing and none of its energy in its mass. In other words, it would have its mass-energy converted into to light-energy, as permitted by Einstein's famous $E = mc^2$ formula.[16]

The computing power of such a device would be formidable. In a ten-millionth of a second it would be able to carry out a calculation that would take a state-of-the-art computer today the age of the Universe to complete. But it would carry out the calculation at a price. If all the available energy is converted into light-energy for computing, a computer would not be anything like a familiar computer. Far from it. It would be a billion-degree ball of light. It would be like a nuclear fireball, a blindingly bright piece of the big bang. Though it might be nice to have the most powerful computer imaginable on your desk, it might be just a little inconvenient.

10:

THE INVENTION OF TIME TRAVEL

Money

Money is not metal. It is trust inscribed.
NIALL FERGUSON, *The Ascent of Money*

If people understood how money was created there would be a revolution before breakfast.
HENRY FORD

'When I was young I thought that money was the most important thing in life,' wrote Oscar Wilde. 'Now that I am old I know that it is.' Money, as Wilde was certainly not the first or last to realise, makes the world go round. But what exactly is money? And how did it originate?

Most people would say that we use money to trade for goods and services: I give you my money in exchange for your goods. Or, alternatively, you give me your money in exchange for my goods. So, behind the question 'What is money?', lies a deeper, more basic question: 'What is trade?'

Rewind the clock maybe 100,000 years. One of our ancestors catches fish. His neighbour makes hand axes. Both need fish and hand axes. Say, the fisherman catches eight fish in the time it takes the axe-maker to fashion four axes. Now, the fisherman could spend half his time catching four fish and half his time making hand axes. But he is not as skilled or as fast at making hand axes as the axe-maker, so he struggles to make one inferior axe. Similarly, the axe-maker could spend half his time making two hand axes and half his time catching fish. But maybe he is not skilled or fast at catching fish, so he struggles to catch two.

This is when one of the men has a genius idea, which he persuades the other makes perfect sense. 'Rather than each of us doing both things, why don't we both exclusively do the thing we are best at – then *trade* our products?' So they do. And the fisherman swaps four of his eight fish for two of the axe-maker's hand axes. He therefore ends up with four fish and two hand axes, which is better than the four fish and one hand axe he would have if he had fished *and* made hand axes. At the same time, the axe-maker ends up with four fish and two hand axes, which is better than the two fish and two hand axes he would have if he too had made hand axes *and* fished.

It seems like a miracle. Both have benefited. And all because of a simple act: trading.

Of course, the fisherman and the axe-maker could have agreed on another exchange rate which was equally advantageous to both. And, even if the fisherman ends up with the same number of hand axes as when he makes them himself, they are likely to be of superior quality. Similarly, if the axe-maker ends up with the same number of fish as when he catches them himself, they are likely to be bigger and tastier, coming as they do from a better, more experienced fisherman.

Trading in this way clearly relies on both parties being honourable. The axe-maker could renege on the deal, taking the fish but not handing over the agreed hand axes. However, if the fisherman and axe-maker belong to the same group, or tribe, there may be an existing template for honourable trade. After all, men, being stronger, may hunt for meat, which they exchange, or trade, for fruit and berries, collected by women. Also, if women move to other tribes to find mates, this may gradually widen the trading circle. Although there is no way to rule out the possibility of cheating, there might be strong incentives not to cheat, which outweigh any tendency to double-cross.

The trading between the fisherman and the axe-maker is limited, however, because it requires two people to meet face to face. An obvious way to boost opportunities for trade is to get a lot of people with a lot of tradable goods together in one place. Such an innovation is a marketplace. If there are several fishermen and several axe-makers (not to mention bead-makers, fur suppliers, fruit collectors, and so on), then they might meet at some place at regular intervals to trade, with the exchange rate being set by supply and demand. For instance, if fish are scarce and a lot of axe-makers want them, fishermen will trade with the axe-maker prepared to offer the most hand axes. Alternatively, if fish are common, fishermen, in order to attract buyers, will have to swap a lot of fish for other commodities.

But trading is more than simply the swapping of goods. It is the swapping of goods between people who have *specialised*: between a fisherman, who has *specialised in fishing*, and an axe-maker who has *specialised in axe-making*. Without such specialisation, there can be no trading for mutual benefit.

And trading, it turns out, encourages ever more specialisation. The fisherman has an incentive to make better fish hooks in order to catch fish more effectively. If his forte is actually locating fish, it may make more sense for him to concentrate on this, opening up an opportunity for someone else to specialise as a maker of fish hooks. Or fishing nets.

Trading and specialisation create more specialisation, which creates more trading opportunities, and more specialisation ... The two feed on each other in a runaway process that, once started, has a kind of unstoppable momentum of its own. Just as evolution by natural selection has created the biological world around us, the idea of trade and specialisation has transformed the human world, creating the commercial world we live in.

Just look around. Pretty much everyone today has a job – a specialised thing that they do. Everyone trades his or her work for other goods and services supplied by other people, who specialise in different things – people who grow avocados or make light bulbs or supply electricity. In fact, trade and specialisation, feeding off each other in an orgy of mutual reinforcement, have in our world proliferated to an extraordinary, mind-blowing extreme. There are hardly any of us that do not use the specialised work of thousands – perhaps millions – of people, most of whom we have never met, across the length and breadth of the world.

The idea of trade plus specialisation appears to be unique to humans. No doubt we will discover that apes do it too but, if they do, it is to a far more limited degree – after all, it is we who have transformed the world not them. Of course, the social insects – ants and bees – have specialised and traded with each other for hundreds of millions of years (there really is nothing new under the Sun). But their societies are frozen into a limited number of castes that perform particular tasks. Humans are uniquely flexible. Given an education and the opportunity, a human can train to be a vet or an airline pilot or school teacher.

But specialisation and trading alone have not created today's commercial world. There have been many other innovations, many other milestones, along the road. Each has boosted trade and accelerated specialisation. And, arguably, the most important is money.

Money, money, money

One of the problems with straightforward trading is that it has to be done *now*. The fisherman has to trade his fish quickly because, in a day or so, they may have gone rotten. But what if the fisherman would rather trade his fish for furs, and the fur trader is not expected at the marketplace for several weeks?

Once upon a time, and it was certainly many thousands of years ago, someone came up with another genius idea. 'I'll give you this token for those fish and then, at any time in the future, you can swap it for the fish equivalent of fruit or furs, or whatever.' The token was of course money. At a stroke, it multiplied the possibilities of trade. Money permitted trade to *time travel*. It was as if someone had invented a Tardis so traders could travel to the future and exchange their goods there. Economists, in less colourful terms, say money is a 'store of value'.

Another problem with straightforward trading is that, to engage in it, two or more people must be physically in the same space – the marketplace. But, say the commodity someone has to trade is bulky and heavy – for instance, a stone for grinding corn – and what they really want to trade it for is beads. However, the market where beads are available is a day's journey over the mountains. Once upon a time, someone had a genius idea. 'I'll give you this token for that grinding stone and, then, anywhere else you go, you can exchange it for the stone equivalent of beads or pots, or anything.' The token, once again, was money. And the innovation multiplied the possibilities of trade. Money permitted trade to *travel through space*. It was as if people had access to a *Star Trek* transporter so they could travel to faraway places and trade their goods there. Once again, economists put it in less colourful terms. They say money is a 'medium of exchange'.

But the genius of money is that not only does it liberate trade in both *time* and *space*, multiplying the opportunities for trade, it has other beneficial properties. Say you do some work for someone who promises that, in a month's time when you finish, they will pay you in a particular commodity – say copper or wheat. It might sound reasonable enough. However, the value of such commodities might fluctuate, depending on supply and demand. This means that you will not know in advance exactly what you will get paid, making it difficult to budget.

Money changes things, however. If your payment is in money, you will know in advance exactly what you will get – unless, of course, you are unlucky enough to live in a time of hyper-inflation such as post-First World War Germany. Money, say economists, is a 'standard of value'.

Of course, if money is to be used as a standard of value, it must be something whose supply does not fluctuate much since scarcity of any commodity boosts its value while plenitude depresses it. One of the first forms of money might have been salt because its source was well known and the technology for extracting it created a supply at a relatively constant rate. Roman soldiers, in particular, were paid in salt, or *sal*, which is the origin of our word 'salary'.

Salt has several other properties that should ideally be possessed by money. It should be *portable* so that it can be easily carried about. And it should be divisible. This allows someone to buy something and get change, which they can use at a later time. Think of the dilemma if the exchange rate is four fish for one hand axe but the fisherman has only three fish. With divisible money, the axe-maker can exchange his axe for the three fish, and receive some change, which he can spend later.

Gold, banks and IOUs

Salt, most people would probably agree, is not an ideal form of money. A better currency is provided by gold, in the form of coins. Although money was used as a unit of account for debts from about 3000 BC, the first gold coins appeared in Greece only around 700 BC.

The trouble with gold is that it is heavy, especially if you are rich. However, there is a clever way around this. Say you are at the goldsmith's one day to exchange some goods for gold – a lot of gold – and the goldsmith says, 'I've got an idea. Instead of you humping those heavy bars around with you, I will give you an IOU. You will not physically have the gold but you will know that I am storing it here for you. And, any time you want it, just come back, present the IOU, and I will give you the gold.'

Maybe, when you find some goods to buy, you will indeed go back to the goldsmith, present your IOU, carry off your gold and exchange it for the goods. But, sooner or later, you will realise there is a better, more convenient, way. Simply present the IOU to the person you are trading with. After all, they will know that, if *they* present it to the goldsmith, *they* will get the gold.

The goldsmith, without perhaps knowing it, has transformed himself into a bank – an entity that stores gold and creates a new kind of money – IOUs. If he stores gold and issues IOUs for other clients as well, sooner or later he will realise that it is highly unlikely that everyone will want their gold back at the same time. So he can create more IOUs than he has gold in the bank. However, if he is prudent – unlike many of the modern banks that triggered the 2008 global banking crisis – he will make sure he has enough gold reserves for an unusual event, when a significant fraction of gold-owners want their gold back simultaneously.

But, just as money is multi-faceted, so too are banks. In addition to creating money and guaranteeing its value, banks perform another key function – they match up lenders and borrowers. Lenders are people who have surplus money they do not want to spend just yet, which they put in the bank. Borrowers are people who need money for some venture – maybe to start a business or buy a house. They expect to earn the necessary money over the following months or years. However, they need the money *now*, and so they borrow it.

This might look like bringing money from the future – future earnings – into the present. But, actually, the amount of money for consumption at any time is fixed. If you borrow money for a mortgage on a house and have to pay it back and so have less to consume, the bank collects your money and lends it to others. So others get to consume instead of you. There is no net transfer of money.

The bank charges the borrower money on top of what he or she borrows because the bank runs the risks of the borrower not paying back the money, or defaulting, and because the bank is a business that needs to make a return, or profit. Some of this interest is passed on to the lenders, so as to make it attractive for them to put their money in the bank in the first place.

To understand what an innovation a bank is, consider what happened before. If you wanted to fund a venture – say, take a ship to the East Indies and trade for spices – you would have to find a very rich backer. There would be a limited number of such people. And a backer might be hesitant to support you since he or she would be risking a lot.

Contrast this with the situation after the birth of banks. To fund your venture, you go to a bank, of which there are many. Because they have combined the resources of a large number of lenders, not only do they have the resources to fund lots of ventures such as yours, but the risk has also been pooled. Each individual lender has less to lose than a single big investor. And, anyhow, the bank can absorb some failures, knowing the majority of ventures have a good chance of succeeding.

A surprising, even shocking, feature of banks is that they never lend out the money that people have deposited. They hold it as a reserve against losses, and for day-to-day cash transactions. Instead, banks *create* money. 'Money comes into existence in the very act of borrowing it,' says economist John Médaille.[1]

Compare the situation of the farmer and the banker. 'The farmer may increase his wealth only by work, the hard work of growing corn,' says Médaille. 'The banker may increase his wealth, or at least his assets, by pressing a few buttons on the computer.' Disturbingly, some banks caught up in the 2008 financial crisis had lent almost thirty times more money than people had deposited with them. By comparison, banks in the nineteenth century lent on average less than five times their deposits.

To reduce their risks when lending money, banks use credit-rating agencies. The banks need to know that you will be likely to pay back what you have borrowed. In fact, credit-rating agencies – and a whole lot of other unseen infrastructure that protects traders – are needed to oil the machinery of commerce. When one person trades with a second person in another country, for instance, he or she needs to know that the second person has the resources to pay and will not simply take the goods and run. The use of credit cards illustrates this well.

Credit cards are a form of short-term credit. In effect, someone takes out an ultra-short-term loan to buy some goods, a loan that is typically paid back in less than a month. Credit cards can be used anywhere in the world to buy goods. And the seller accepts a card as payment because he or she trusts that the issuer of the card checked that the card-holder had the resources to pay. Even if the issuer got it wrong, the seller knows that the issuer, who has shouldered the risk, will guarantee the payment.

Booms and busts

In addition to banks, money and the rest, there have been many more milestones on the road to the commercial world we live in, each of which, to a lesser or greater degree, has boosted the number and frequency of trades. Many people still live in poverty. But the result of the frenzy of trading over the millennia has been steadily to increase the standard of living of the average person. Today, a typical home in the First World boasts a variety of goods that a couple of centuries ago would have been owned only by a king.

However, today's standard of living has not been brought about solely by an ever more extensive web of global trading. Hand in hand with the expansion of trade has been a rise in the amount of energy consumed per person. Once upon a time, a person had no choice but to use his or her own labour or – as too often happened – the labour of other people as slaves. Later, with the domestication of horses, a person had access to the energy equivalent of maybe ten people. This was later magnified by technological innovations such as windmills and water-powered factories. But the most significant boost in the energy available per person came with the utilisation of fossil fuels such as coal. These were resources laid down hundreds of millions years ago – trees, which had soaked up sunlight, died, and became buried and compressed deep in the Earth. This enabled each person to have access to the energy equivalent of maybe 150 people. Whereas windmills and water-powered mills exploited today's sunlight – which of course drives the wind and evaporates water, which falls as rain – coal brought into the present-day economy the resource of *yesterday's sunlight*.

So it is debatable whether it is trade or the exploitation of ever more potent energy sources that is responsible for the rise in the standard of living of the average person over past millennia. Trade makes it commercially viable to mine coal and develop nuclear power stations. And the energy made available boosts trade.

But all is not rosy in the financial garden. Not only do a large number of the people in the world remain in poverty but the global financial system has a tendency to lurch from boom to bust. The reasons for this are much debated. Certainly, in the boom times, people have a tendency to become over-optimistic and borrow more than they should. So, when the bust comes, they have debts to pay, which starves the recovering economy of any money they might invest. It means that the peaks of the financial cycle are steeper and the troughs deeper than they would otherwise be. But, although this exacerbates busts, it does not explain the *reason* for the booms and busts.

One possibility often touted is that they are caused by shocks from outside. Everything is chugging along nicely, goes the story, then along comes a technological innovation that throws a spanner in the works – for instance, the World Wide Web, which triggered the infamous dot-com boom, followed inevitably by the dot-com bust. Another possible reason for booms and busts is that investors put the money into the construction of factories to supply certain goods. The factories employ many people. But eventually there are so many factories making the goods that they supply more than is needed. Because of this supply over-shoot, people lose their jobs. Ultimately, it is all down to the fact that, while investment sky-rockets, what people consume very definitely does not.

Another possible reason for booms and busts is that, when demand for goods slackens off, the suppliers of those goods do not react by reducing prices. One reason for this is that they know how much resistance there will be to cutting wages. By cutting wages and the prices of goods, demand might have been stimulated. Instead, demand falls and people lose their jobs.

Booms and busts, despite the periodic claims of economists, appear to be uncontrollable. The question then arises: do we really understand the complex, multiply connected commercial world we have created? The answer, worryingly, appears to be no. As the Canadian economist John Kenneth Galbraith said, 'Economics is extremely useful as a form of employment for economists.'

Nick Leeson is infamous for sinking Barings, Britain's oldest merchant bank, in 1995. In an interview on 30 September 2012, *The Sunday Times* asked the rogue trader, 'What's the most important lesson you have learned about money?' His answer? 'None of us ever knows enough about it.'

II:

THE GREAT TRANSFORMATION
Capitalism

Capitalism is the astounding belief that the most wickedest of men will do the most wickedest of things for the greatest good of everyone.

JOHN MAYNARD KEYNES

Under capitalism, man exploits man. Under communism, it's just the opposite.

JOHN KENNETH GALBRAITH

Capitalism is fascism minus murder, the American author Upton Sinclair once implied. Perhaps a little extreme. But it does throw into focus the question: what is capitalism? Bizarrely, despite it being a system of commerce in which most people in the world participate, few ever stop to think about what it is and how to define it.

Not surprisingly, perhaps, the central plank of capitalism is capital – goods, land, factories, and so on – the kind of things Karl Marx dubbed the 'means of production'. In capitalism, people are free to own capital. This might not seem a significant freedom because we take it so for granted. However, contrast capitalism with communism, in which private property is forbidden, or medieval feudalism, in which only an elite had the rights of ownership.

The freedom to own capital, however, is only half the recipe for capitalism. The other ingredient is the freedom to trade that capital for profit.

How capitalism works in practice is complex. But highlighting some of the common myths of capitalism helps to shine a light on its inner workings.

Global capitalism has spawned an extensive web of trades called the market. The market automatically matches up the supply of goods with the demand for those goods in the optimum way possible – or at least that is the theory. To operate effectively, such a market should be free – that is, unfettered by political regulation or regulation of any kind. This is the mantra of an influential group of people who advocate free-market capitalism as a panacea for all ills. The laissez-faire idea has its origins with Adam Smith, the Scottish philosopher and economist who, in 1776, published one of the most influential books in the history of thought. In *The Wealth of Nations* – or, to quote its full title, *An Inquiry into the Nature and Causes of the Wealth of Nations* – Smith argued that the free market was the best model for the economy.

The myths of capitalism

Even Smith recognised that a true free market is a myth. It would be deemed unacceptable by the overwhelming majority of people. A free market would, for instance, permit a trade in anything and everything – including child labour. Once upon a time such a trade did indeed exist in countries such as Britain. But, nowadays, it is pretty much universally agreed that child labour is unacceptable, and stringent regulations are put in place to prevent it.

In addition to regulations controlling what kind of labour may be traded, there are also regulations that severely limit or forbid the trade in goods deemed dangerous to society such as heroin and plutonium. And it is not only the goods that can be traded that are regulated, so too are the companies that are permitted to trade. In order for a company sell its shares, for instance, it must be listed on a stock market. But, before this is possible, there is a rigorous vetting process that might last up to five years. Even when finally listed, a company is permitted to sell shares only to certified traders. And, if the share price drops below a certain level, trading in the shares might be suspended for a market holiday.

'Market forces don't exist in a vacuum,' says Joseph Stiglitz, author of *The Price of Inequality*. 'We shape them.'[1] 'All markets are not only constructed and regulated but constantly manipulated,' says Ja-Hoon Chang, author of *23 Things They Don't Tell You About Capitalism*.

Perhaps the most striking way in which the market is straitjacketed and defined by political decisions is seen in the wages of workers. These are wildly different for comparable jobs in different countries. For instance, a taxi driver in London might be paid about thirty times more than one in Dhaka, Bangladesh, in terms of the goods he can buy locally. If a free market existed, the wages of taxi drivers would all be roughly the same. After all, if a taxi driver in Dhaka was unhappy with his level of pay, he would simply relocate to London, where he would be paid a lot more. With taxi drivers shuttling back and forth around the globe in search of better pay like this, sooner or later the wages of all taxi drivers would equalise.

However, taxi drivers cannot easily up sticks and move around the globe because most countries have erected high and impenetrable barriers to immigration. This is the principal reason for why those wages are so wildly different in different countries.

If something as fundamental as the reward that people receive for their work is determined by political decisions rather than market forces, then the free market must be a mythical beast. Everywhere, the market is tightly constrained by regulations imposed from outside. Nowhere is it free – nor would any civilised country allow it to be free.

Consequences of the free market myth

The idea that a free market will optimally match the supply of goods with demand has consequences. Those who believe in the idea push for deregulation of more and more of the economy. If the market does not work in delivering what they hope for, they say the reason is because the market is not free enough, and push for yet more deregulation. However, such deregulation is widely believed to have triggered the serious global financial crisis of 2008 when big financial institutions such as banks embraced very risky investments.[2]

But a free market not only permits risky behaviour that can threaten the global economy itself. It can have a more direct and malign effect if imposed on developing countries.

If such a country wants to create, say, a motorbike industry, then to get things going it might subsidise motorbike manufacture while, simultaneously, protecting its fledgling industry by imposing tariffs on imported motorbikes. This will often provoke developed countries to complain to the World Trade Organisation that the country is preventing a free market in motorbikes. The WTO can then authorise trading partners to impose sanctions on the developing country unless it removes its subsidies and its protective measures. It of course does.

The problem is that, when an industry is in its infancy, it is nowhere near as efficient as a mature industry. Also, its identity, or brand, is not well known or respected. Consequently, the motorbikes being built by the developing country cannot compete in either quality or price with imports, and the industry dies. Contrary to the free marketeers' mantra, the free market has not made a poor country richer. At best, it has left it to stagnate.

The big irony here is that the rich countries of the world all got rich by *both* subsidising and protecting their fledgling industries over periods of many decades. For instance, nineteenth-century Britain, by imposing prohibitive tariffs on imported Indian textiles, effectively destroyed the Indian textile industry. It was then able to sell the products of its own textile industry to the subcontinent. Earlier, in the eighteenth century, Britain had used protection measures in order to catch up with the Netherlands. And, in the nineteenth century, the US used exactly the same tactic to catch up with Britain. 'In forcing a free market on developing countries, the rich countries seem to have forgotten their own histories,' says Chang. 'Not surprisingly, those in developing countries find the double standards deeply annoying.'[3]

The invisible hand

Behind the faith in the mythical free market is the belief that such a market will optimally match up supply and demand. Market advocates talk of this being achieved by the 'invisible hand'. No one is able to define it but everyone claims it is working behind the scenes. Alan Greenspan, former chairman of the US's Federal Reserve Board, has even stated that the market is simply 'too complex for anyone to understand'.

Simultaneously promoting the market and believing it to be too complex to understand is to trust the lives of billions of people to an unpredictable system. The market has undoubtedly played a key role in delivering a rising average standard of living to much of the world's population over the past few centuries. However, it has also delivered massive environmental problems of pollution, habitat destruction and, most seriously, global warming caused by the burning of fossil fuels such as coal and oil. To say that the market is too complex to understand is to accept that the fate of the human race – global warming threatens to extinguish human life, though not life itself – is out of our hands, that we are at the mercy of the caprices of fate. 'The conservative argument is that the economy is like the weather, that it just operates automatically,' says Sidney Blumenthal, former adviser to President Bill Clinton.

Believing that the market is too complex to understand is a seductive idea. Economists and politicians need not worry about the hard problem of how exactly the market works – or does not. However, this is a cop-out, according to Chang. 'No matter how complex it is, we must try hard to understand it in order to ameliorate its ill effects,' he says. 'The market, after all, is an entirely human creation.'[4]

Earthquakes and markets

Almost without exception, economists since Adam Smith have considered that the free market is in a state of equilibrium, naturally balancing people's conflicting aims and desires. This, however, rather contradicts reality. Every decade since 1776, there has been a crash, bust, downturn, depression or slump.

Economists often say that such events are merely unusual external shocks to a market. But there are so many of them that such an explanation is wearing a little thin. 'The pronounced frequency of market upheavals is precisely what is most constant in economics,' says physicist Mark Buchanan.[5]

There is now a growing belief that the current economic theory – which fails to predict crashes, the most striking feature

of the market – is inadequate. Many believe that a better economic theory is required that encapsulates the inherent instability
of the market. This was actually a view advanced in the 1930s
and 1940s by American economists such as Irving Fischer. But it
was buried by Milton Friedman and his allies in the Chicago
School, who vigorously promoted the idea of equilibrium in
markets. 'A well-designed policy must begin with a situation
approximating that which actually exists,' warned Nobel Laureate
Ronald Coase. 'The situation that exists in any real-world market
is one rife with "market failures".'[6]

The market has a peculiar and counter-intuitive feature. Large
fluctuations in prices are more common than expected. Our
intuition of everyday life, for instance, tells us that most men
weigh between 9 and 13 stone, a few weigh 20 stone, and a very
few 40 stone. This conforms to a so-called Gaussian distribution,
more commonly known as a bell curve because of its similarity
with a bell jar. An upturned bell jar is fat in the middle and tapers
off far from its centre. This is exactly the shape you would see if
you took a piece of graph paper and plotted the weight of men
along the horizontal axis and the number of men at each weight
of men up the vertical axis. Like the bell jar, the distribution of
the weights of men clusters around an average and tails off
rapidly far from the average.

Contrast this, however, with fluctuations of the market. In a
typical day, the price of stocks changes by less than 2 per cent.
But it is possible to have a fluctuation of 20 per cent or 50 per
cent. This would be like there being men weighing 10 times the
average. Or 25 times. Mathematicians say that, unlike the bell
curve, the distribution of market fluctuations has a 'fat tail', which
is just a technical way of saying that extreme events are far more

likely than our everyday intuition would predict.[7] 'A credible economic theory of markets – something we do not yet have – would explain why the distribution of market returns shows such a preponderance of large events,' says Buchanan.

In fact, according to Buchanan, market movements conform to a very simple mathematical pattern. Larger movements of, say, 10–15 per cent, are less likely than movements of 3–5 per cent. In fact, the probability of a movement of any size decreases in inverse proportion to the cube of its size. So, if moves of 5 per cent or more have a certain chance of occurring, moves of 10 per cent or more are $2^3 = 8$ times less likely; and moves of 20 per cent or more are another $2^3 = 8$ times less likely.

This striking pattern, which is seen in markets for stocks, foreign exchange and futures, is reminiscent of a whole range of natural phenomena, from solar-flare activity to frequency of mass extinctions to frequency of earthquakes on the San Andreas Fault.[8] 'All these systems and many others exhibit a naturally irregular rhythm in which long periods of relative quiescence are sporadically broken by bursts of intense upheaval,' says Buchanan.

If a financial market is a system in equilibrium it is hard to see why market movements should behave like earthquakes. But it makes more sense if a market is merely another system, like the Earth's crust, which is constantly driven out of balance by various forces, and responds to those forces in complex, dynamic ways.

The new economic models that are being concocted, often by physicists, are driven by instability and feedback, which is the case in many natural systems. 'It's fair to say these models don't yet give us an adequate understanding of the basic patterns we see in markets, but they at least move in the right direction by taking the historical data seriously and trying to explain it,' says Buchanan. 'Nothing in mainstream economics seems as likely to succeed in this.'

Some of the converts to the new econophysics are calling for a new Manhattan Project for economics. With the well-being of 7 billion plus people depending on the global economy, it would arguably be money well spent.

Gratuitous complexity

But, though the market is complex and ill understood, people are nevertheless adding gratuitous complexity to the market. A good example is the creation of complex financial instruments, which played a significant role in triggering the global financial crisis of 2008. One of the most dangerous, according to Chang, was the Collateralised Debt Obligation, or CDO. This bundles together thousands of risky investments such as loans for the purchase of houses by people with little prospect of being able to pay them back. To make this 'Voodoo process' even more risky, says Chang, CDOs themselves have been bundled together in their thousands to create an even more complex product known as a CDO-squared. 'The level of unnecessary complexity is staggering,' says Chang. 'To understand such an investment typically requires reading in excess of 1 billion pages of documentation.'

Greenspan has said that such products are 'too complex to regulate' because nobody understands them. In fact, says Chang, they are designed to be opaque, to obscure. 'Then, if they go wrong, it is impossible to determine why or to determine who is responsible.' This was particularly true of the 2008 crisis, in which hardly anybody has put up their hands and accepted blame.

Chang believes it is irresponsible and dangerous to introduce gratuitous complexity into the market. He makes a comparison with drugs, which the pharmaceutical industry must rigorously test, often for many years, before they can be safely released into the world. He believes that there should be a similar rigorous testing of financial instruments or any schemes that increase complexity of the market. Only then would we be confident they would not destabilise the market.

The biggest market myth

But, actually, the malaise with the market system might go deeper than the complex financial instruments that played such a role in destabilising everything in 2008. Proponents of the market system talk of it being a natural state, the economic analogue of the web of life created by Darwin's theory of evolution by natural selection. But this, according to the Hungarian economist Karl Polanyi, is the biggest myth of all. In his influential book, *The Great Transformation*, he made the case that the market system is a relatively recent innovation.[9]

The laws and regulations that made it possible were introduced by the governments of modern states beginning in the late eighteenth century. Laissez-faire was planned. It was an integral part of the industrial revolution. Before this time, claimed Polanyi, many financial transactions were as much about gaining social status or reinforcing social bonds as about making money. 'Land, labor and money itself were not regarded chiefly as commodities to be bought and sold,' writes David Bollier, author of *Silent Theft*. 'They were "embedded" in social relationships.' In other words, people were motivated by the interests of a family, clan, or village rather than by mere self-interest, the defining feature of a market economy.

According to Polanyi, the Great Transformation turned land, labour and money into abstract commodities and encouraged the belief that they could be exploited without limit to deliver eternal growth. Such growth is measured by Gross Domestic Product, which is the monetary value of the production of all goods and services in an economy. However, if an economy is continually to grow, people must continually spend, which is possible – for a while – only if people borrow money so their debts accumulate on a massive scale. Perpetual growth is of course unsustainable. And the consequences of trying to achieve it are serious. 'Because markets treat nature as essentially limitless and human beings as commodities, they are always pushing human societies and nature to the breaking point,' says Bollier. This pursuit of profit without concern for the consequences has brought the planet to the edge of ecological disaster.

According to Polanyi, the Great Transformation also encouraged the belief that the free market is pure and divorced from society, which ignores the fact that labour involves people, who have lives and loves, hopes and dreams. In 2013, across Europe and America, human talent and industry is being wasted on a huge scale as governments employ austerity measures in an attempt to pay for the excesses of the market system. 'We need to enlarge the scope of political conversation to include such questions as: how shall we re-integrate market forces into society so that they can be constructive and not disruptive?' says Bollier.[10]

'Nowadays people know the price of everything and the value of nothing,' Oscar Wilde observed in the nineteenth century.[11] A common refrain in the twenty-first century is 'Capitalism killed

Communism. Now it's coming for Democracy.' But human social behaviour is motivated by more than profit and self-interest. 'I shop therefore I am' does not define a human being. Adam Smith envisioned the free market as the servant of humanity not its slave master. Patrick McGoohan, playing the lead role in the cult 1960s TV series *The Prisoner*, could have been speaking for millions of ordinary people today when, in exasperation, he cried out, 'I am not a number! I am a free man!'

PART THREE: Earth works

NO VESTIGE OF A BEGINNING

Geology

Rocks are records of events that took place at the
time they formed. They are books. They have a
different vocabulary, a different alphabet, but you
learn how to read them.

JOHN MCPHEE

Civilisation exists by geological consent, subject
to change without notice.

WILL DURANT, *Ladies Home Journal*, January 1946

The world has not always been the way it is. This is one of the most powerful and revolutionary insights in history. It spawned a new science – geology – and it inspired Charles Darwin to recognise that all creatures on Earth have diverged from a common ancient ancestor.

The evidence that the Earth is not static – that it was not made in its current form by a Creator – is subtle. For instance, on Madeira, a volcanic island off the north-west coast of Africa, fossil seashells are commonly found more than 6,000 feet up on the summit of the tallest mountain. How did they get there? The obvious but mind-blowing answer is that the mountain began its life beneath the sea and rose skyward.

Mountains do not rise by a noticeable amount in a human lifetime. Consequently, it must have taken a huge number of human lifetimes for Madeira's tallest mountain to have risen from beneath the sea to a height of more than 6,000 feet.

'A huge number of human lifetimes' is not exactly precise. Fortunately, other subtle evidence exists that can provide the precision. Scientists in the eighteenth century could see mud accumulating at the bottom of lakes, deposited there by the rivers and streams. They could also see cliffs and other exposed rocks that looked remarkably like mud, piled thin layer upon thin layer. The suspicion grew in their minds that the rocks had been made by mud settling to the bottom of an ancient body of water. Such a process was extremely slow – in a century, it would deposit no more than a fraction of an inch of mud. The unavoidable conclusion was therefore that the rocks must be hundreds of millions of years old – created over *millions* of human generations.

For the first time in history humans contemplated Deep Time, compared with which their existence was as transient as that of a firefly in the night. The Earth is not just old, it is *beyond-human-comprehension old*. 'There is no vestige of a beginning, no prospect of an end,' said Scottish scientist James Hutton in 1788.[1] Today, we know from the radioactive dating of meteorites, the builders' rubble left over from the formation of the Solar System, that the Earth is about 4.55 billion years old – that is 4.55 *thousand* million years.

Mud becomes mudstone after it is deposited on the bed of a body of water, then compressed by the layers of mud deposited on top. The creation of such sedimentary rock illustrates another profound insight.[2] The past, contrary to novelist L. P. Hartley's famous opening sentence, is not a foreign country; they do not do things differently there.[3] The processes that have changed the Earth's surface are nothing more than the processes that are going on today – weathering, volcanic eruptions, and erosion by water and wind. Working away over mind-cringing spans of time, they can literally move mountains – or grind them into microscopic dust.

Two mountain ranges that illustrate this are the Himalayas, which today are rising skyward, and the Caledonian mountains of Scotland, the eroded stumps of a Himalaya-like chain born about 500 million years ago. Both ranges have been created by an identical process – the titanic collision of giant chunks of the Earth's crust. The evidence for this can be seen in both locations in the form of huge folds created by layers of colliding strata rucking up over each other.

The fact that chunks of the crust can collide like this leads to another revolutionary insight. Although the early geologists believed that the surface of the Earth merely moved up and down, creating features such as mountains, in fact, the surface also moves *sideways*.

Plate tectonics

As far back as 1620, Francis Bacon, poring over the first semi-accurate maps of the world, noticed a remarkable similarity between the coastlines of Africa and South America. Like two giant jigsaw pieces, they appeared to fit together. This was considered no more than a curiosity until the early twentieth century. Then, a German geologist suggested an idea so controversial that he would die unrecognised, having convinced essentially no one of its truth.

Alfred Wegener's extraordinary idea was that the continents *move*. The reason the coastlines of South America and Africa fit is that long ago they were *joined*. They then split and drifted apart. Wegener's evidence of such continental drift was that not only are the rocks on either side of the join the same but so too are the fossils.

The main reason no one believed Wegener's idea was that he could provide no mechanism for continental drift. Also, South America and Africa were separated by thousands of kilometres of seabed. How in the world could they have crossed such a substantial and solid barrier?

What changed everything was the surveying of the seabed of the Atlantic. The laying of transatlantic telephone cables had revealed a curious ridge in the mid-Atlantic.[4] Sonar surveys by the US Navy in the 1960s revealed it was more than a ridge. It was a stupendous mountain range that bisected the Atlantic, stretching 10,000 kilometres from Iceland down to the Falklands. What was it doing there?

A key piece of evidence came from measurements of the magnetic field of the rocks on the seabed. Those rocks were originally spewed out as lava by ancient volcanoes. When the lava was liquid, its atoms were free to align along the direction of the Earth's north–south magnetic field of force; when the lava solidified, the atoms froze for all eternity in the direction of the ancient field.

The magnetism of the seabed rocks revealed an extraordinary pattern. On either side of the mid-Atlantic Ridge were symmetric stripes of magnetism: first rocks were magnetised in one direction, then in the opposite direction, over and over again. What did it mean?

Actually, measurements of the magnetism of rocks on the land had already shown such magnetic reversals. The Earth's magnetic field is pretty much like that of a bar magnet, and at intervals it flips direction. What was the north magnetic pole becomes the south magnetic pole, and vice versa. To this day, nobody is quite sure why this happens. But the stripes of rock, magnetised first one way, then the other, gave the geologists of the 1960s a powerful tool. Dating the rock of the stripes showed that the oldest were furthest from the mid-Atlantic Ridge while the youngest were closest.

Suddenly, it became clear what was happening. The mid-Atlantic Ridge was *manufacturing crust*. About 120 million years ago, South America and Africa had indeed been joined at the hip.

Then a huge crack in the Earth had opened, spewing forth lava. Water had flooded in. Year after year, century after century, millennium after millennium, lava had gushed out of the tremendous fissure in the Earth's surface, creating ever more crust, which pushed the two continents further and further apart. Something like this is happening today at Afar in Ethiopia, where not two but *three* chunks of the Earth's crust are pulling apart and a new ocean is being born.

Wegener's critics were wrong to ridicule him. It was not necessary for South America and Africa to cross a vast expanse of solid seabed to reach their current positions. At the outset there had been *no seabed*. It grew between the land masses, in the process pushing them remorselessly apart.

The figures are impressive. The mid-Atlantic Ridge pumps out about 5 cubic kilometres of lava *every year*. This is twenty times as much as Mount St Helens, which in its mighty volcanic eruption of 1980 spewed out a paltry 0.25 cubic kilometre of rock. Mid-ocean ridges – and they are not found only in the Atlantic – are crust factories. Globally, they make about 30 cubic kilometres of new crust every year.

But it is impossible to create ever more crust on a ball such as the Earth which is finite in size. Something must give. And it does.

To understand what happens it is necessary to know one other fact: the crust of the Earth is fractured into about twelve major chunks, or plates. A plate might carry on its back continental crust or oceanic crust or both. Wegener, who was right on so much, was therefore wrong to think it was merely continents that were drifting. The plates float on the mantle, the super-hot, super-dense fluid of the Earth's interior.

A subtlety here turns out to be very important. Oceanic crust, made of volcanic rock, is denser than continental crust, which is made of granite, formed from volcanic magma. Consequently, continental crust floats higher. This explains something we so take for granted that it is never remarked on: the ocean bed is low and wet while the continents are high and dry. Continental crust, as geologists like to say, is the 'scum of the Earth'.

So now it is possible to understand what happens as crust is ceaselessly manufactured at mid-ocean ridges. If two plates carrying continental crust collide, in the stupendous collision the light continental crust rucks up, rising to create mountains. This is happening at the site of the Himalayas. But the rucking up of coast is only a temporary solution to the ever-increasing mass of crust. Somewhere crust must be destroyed. And it is: where a plate carrying oceanic crust runs into one carrying continental crust. This is happening today along the west coast of South America.

Oceanic crust, being denser than continental crust, dives down under it into the mantle. As it does, it carries with it water and seashells and all kinds of oceanic detritus. This is very significant because these things have the effect of lowering the melting point of the continental crust under which the ocean crust is diving. The result is volcanoes on the continental crust above. These can be seen today along the length of Chile. They are the Andes.

Actually, it is not always the case that, when oceanic and continental crusts collide, the oceanic crust dives underneath. In the collision off the west coast of Britain, the oceanic crust is *pushing* the continental crust along with it. This is widening the Atlantic by about 5 centimetres a year. Britain and the United States are in the midst of a long goodbye.

But, when oceanic crust does dive beneath continental crust, it does not dive down smoothly. As it plunges to oblivion in the mantle, it snags, and judders forward. And this juddering creates tremendous earthquakes, such as the one that struck Chile in 2010.

So plates are made and are destroyed. They run into each other and slide past each other. This is happening along the San Andreas Fault in California, where the Pacific Plate is moving past the North American Plate, closing the gap between Los Angeles and San Francisco by 5 centimetres a year.

Plate tectonics explains all we see on our planet. Without it, geology would make as little sense as biology without Darwin's theory of evolution by natural selection, or genetics without DNA.

But what drives the motion of chunks of the Earth's surface? Wegener, who died in 1930, aged only fifty, on a field trip in Greenland, failed to find it. But it is nothing more complicated than heat trying to escape from the interior of the Earth. Incredibly, 4.55 billion years after its creation, the planet still retains some of the heat of its molten birth.[5] This is because, being a big body, it has a relatively small surface area through which heat can escape compared with its volume, and so heat has a hard time getting out. The interior of the Earth is also continually heated by the disintegration, or decay, of radioactive elements in its rocks such as uranium, thorium and potassium.

All this keeps the interior of the Earth fluid (though it is a *very* viscous fluid). And, just like water in a saucepan on a hot-plate, the fluid roils, with hot, light fluid rising and cold, dense fluid sinking. Nobody knows whether this convection occurs in one big circulating cell that extends down to the Earth's core or whether the pattern is more complex. But the basic idea is straightforward. Circulation of fluid in the mantle drives the motion of the plates.

But the plates do not only continually change the shape of the Earth's surface; they also play a key role in keeping our planet habitable.

Carbon dioxide gas is constantly pumped into the atmosphere by volcanoes. It is sucked out of the air by the oceans and finds its way into the carbonate shells of sea creatures. When they die, they settle on the seabed. Their remains are therefore taken down into the mantle when an oceanic plate dives under a continental plate. In this way, the plate tectonic conveyor belt prevents carbon dioxide in the atmosphere building up to dangerous levels. Carbon dioxide is a potent greenhouse gas that traps heat in the atmosphere.[6]

What happens if there are no plate tectonics to remove atmospheric carbon dioxide is apparent on Venus. Carbon dioxide from volcanoes has built up to such levels that the planet has an atmosphere about 92 times thicker, composed solely of the gas. It makes the surface hot enough to melt lead.[7]

The Earth's plates, as they dive down into the mantle, might appear to be forever beyond our view. However, seismic waves bouncing around inside the Earth from earthquakes can be used, with the aid of computer wizardry, to create a kind of X-ray of the interior of the Earth. Such seismic tomography shows the skin of the crust wrapped around the mantle. Deep down is the core. This consists of an outer core of liquid iron wrapped around an inner core of solid iron. Remarkably, seismic tomography appears to show whole slabs of plates sinking down to the core. This is surprising since they would be expected to melt. The slabs pile up outside the outer core.

If indeed the core is a plate graveyard, it could explain another phenomenon. Plumes of superhot mantle appear to rise from the core. They heat the underside of a plate like a blowtorch. There is such a superplume under the Hawaiian chain of islands. In fact, each island is a volcano born as the plate drifted across the blowtorch.

The core is at a temperature of about 5,000 °C, comparable to that of the surface of the Sun. It could be that the plate grave-yard on the outer core permits heat from the core to escape only where there are gaps in the piles of plates, and that this is the origin of the superplumes.

Though we can never go there, the interior of the Earth is gradually yielding its secrets. 'The world is the geologist's great puzzle-box,' said Swiss geologist Louis Agassiz in 1856. 'He stands before it like the child to whom the separate pieces of his puzzle remain a mystery till he detects their relation and sees where they fit, and then his fragments grow at once into a connected picture beneath his hand.'[8]

13:

EARTH'S AURA

The Atmosphere

The sun, moving as it does, sets up processes of change and becoming and decay, and by its agency the finest and sweetest water is every day carried up and is dissolved into vapour and rises to the upper region, where it is condensed again by the cold and so returns to the earth.

ARISTOTLE, *Meteorology*, 350 BC

It's raining men! Hallelujah! It's raining men!

THE WEATHER GIRLS

'The thickness of the Earth's atmosphere, compared with the size of the Earth, is in about the same ratio as the thickness of a coat of shellac on a schoolroom globe is to the diameter of the globe,' said Carl Sagan.[1] Yet this insignificant sliver of haze makes life on our planet possible. Not only does it act like a blanket, trapping precious warmth, it evens out wild extremes in temperature between night and day. Without the atmosphere, the Blue Planet would not be blue. It would be a dazzlingly white ball of ice with an average temperature of -18 °C.

About 4.55 billion years ago, when the Earth was newly minted, the atmosphere is believed to have been made mostly of carbon dioxide, spewed from volcanoes. But about 3.8 billion years ago, the planet came under violent and sustained bombardment by city-sized asteroids. This Late Heavy Bombardment not only turned the surface molten but blasted into space the early atmosphere and all the water.[2] Evidence points to icy comets later bringing much of the water that today covers 71 per cent of the planet's surface.[3]

Today's atmosphere consists of about one-fifth oxygen and four-fifths nitrogen, with a few trace gases such as argon, water vapour and carbon dioxide. In marked contrast with its primordial antecedent, it is almost entirely the creation of life. For aeons, blue-green algae, or cyanobacteria, pumped oxygen – the waste product of photosynthesis – into the air. This combined with the planet's plentiful iron to make iron oxide, creating tremendous deposits of reddish-brown rocks, which can be seen in Australia today. When the iron could soak up no more, oxygen built up to catastrophic levels in the atmosphere. It poisoned large numbers of organisms. Crucially, however, oxygen provided the super-charged energy source for animals and, one day, humans.[4]

Circulation

But the atmosphere is more than an oxygen-rich blanket that shrouds the world. It is a layer of air in ceaseless motion, driven by solar energy. The Sun heats the equatorial regions more than it does the poles, making the equator hotter than the poles.[5] Since heat always flows from a hot body to a cold body in order to even out the temperature, heat flows from the equator to the poles through the atmosphere in an attempt to iron out the global temperature. 'The Earth and its atmosphere constitute a vast distilling apparatus in which the equatorial ocean plays the part of the boiler, and the chill regions of the poles the part of the condenser,' wrote the nineteenth-century English physicist John Tyndall.[6]

If the Earth was not spinning – or it was spinning very slowly, like Venus – it would be particularly simple for heat to travel from the equator to the poles.[7] Hot air, being lighter than cold air – think of hot-air balloons – would rise at the equator and travel towards the poles.[8] There, it would lose its heat, sink down, returning to the equator closer to the surface. Such a continuous conveyor belt of air is known as a Hadley cell after George Hadley, the English lawyer and meteorologist who proposed it in 1735. In fact, a non-rotating Earth would support *two* Hadley circulation cells – one in the northern hemisphere and one in the southern hemisphere.

The Earth, however, *is* spinning rapidly – once every 24 hours. Consequently, the ground – and therefore the air – is moving most quickly at the equator and most slowly at the poles. This is why NASA, the American space agency, launches from Florida, and ESA, the European Space Agency, from Kourou in French Guyana. By lifting off as close as possible to the equator, rockets get the maximum boost from the Earth's rotation.

Without even knowing it, people at the equator are travelling at almost *twice* the speed of a Boeing 747 – about 1,670 kilometres per hour.[9] The consequence for hot air, travelling away from the equator, is that it finds itself continually travelling faster than the ground below it. From the point of view of someone on that ground, the air therefore appears deflected in the direction of the Earth's rotation – to the right, or *east*, in the northern hemisphere and to the left, or *west*, in the southern hemisphere.[10]

So extreme is the deflection of north- and south-travelling air that there is no simple way for heat to get from the equator to the poles. The circulation, instead of forming a simple Hadley circulation cell, splits into *three* – in other words, in each hemisphere, there are *three* overturning cells of air. Think of them as three parallel bands, each spanning about a third of the distance between the equator and the pole.

On a faster-spinning planet, the circulation splits into *even more* bands. Jupiter, for instance, rotates in a mere 10 hours despite having an equatorial diameter about 11 times that of the Earth. This super-fast spin – the planet's equator is moving about 25 times faster than the Earth's – causes its circulation to split into about 15 bands – 7 on each side of the equator band.

While the lighter bands are called zones, the darker bands are called belts.

In the circulation band in the polar regions of our planet, relatively warm air moves towards the pole at a high altitude. Because it finds itself moving faster than the ground, it appears to someone on the ground to be deflected in the direction of the Earth's rotation. Consequently, the high-altitude winds are westerly; they blow *from* the west, in the same sense as the Earth's rotation. When the air reaches the poles, it cools and sinks. It then returns, at a lower altitude, whence it came. Because it now finds itself moving more slowly than the ground, from the point of view of someone on the ground, it appears deflected in the direction opposite to that of the Earth's rotation. This means that the winds close to the ground near the polar caps blow mainly easterly – that is, *from* the east, which is against the sense of the Earth's rotation.

Something very similar happens in the closest of the circulation bands to the equator. High-altitude air flowing away from the equator appears from the ground to be deflected in the direction of the Earth's rotation. Such winds in the tropics therefore blow *westerly*. Low-altitude air flowing back to the equator, on the other hand, appears to an observer on the ground to be deflected against the Earth's rotation. This is why the winds near sea level in the tropics – known as the trade winds – blow primarily easterly.

Weather

The most interesting of the three terrestrial circulation bands, however, is the middle one, halfway between the polar band and the tropical band. Here, at mid-latitudes,[11] the Earth's rotation has its biggest effect on the air moving north and south.[12] This makes the circulation inherently unstable, leading to the constant spawning of eddies, or vortices. The middle circulation cell is also the domain of super-fast westerly winds at high altitude. This jet stream can blow at more than 400 kilometres per hour and steers weather systems. It is why flying from America to Europe is quicker than flying in the opposite direction.

This is a good place to dispel an old wives' tale. People often say water swirls down a plughole consistently one way in the northern hemisphere and the other way in the southern hemisphere. It does not. The water can swirl either way, depending on the initial oomph it gets – from the flow from the tap or from any unevenness in the sink itself. Differences in the speed of the Earth's surface due to the planet's rotation are simply too small across the tiny span of sink to have any effect on the water. But this is not true for an air mass that is hundreds or more kilometres across. In marked contrast with water swirling in a sink, these *do* indeed spin differently in the northern and southern hemispheres.

It works in this way. Imagine a region of low pressure, known as a cyclone, in the northern hemisphere.[13] Surrounding air rushes in from all sides to try to equalise the pressure. Air rushing in in a northerly direction finds itself moving faster than the ground – that is, from the ground it appears to be deflected eastward, in the direction of the Earth's spin. Air rushing in in a southerly direction finds itself moving more slowly than the ground – that is, it appears deflected westward, in an opposite sense to the Earth's spin. The effect of this is to spin the air mass anticlockwise. (In the southern hemisphere, a cyclone spins clockwise.) For a high-pressure system, known as an anticyclone, the opposite reasoning applies. An anticyclone spins clockwise in the northern hemisphere and anticlockwise in the southern hemisphere.

Weather is loosely defined as 'day-to-day variations in atmospheric conditions'. It occurs in the lowest layer of the atmosphere, or troposphere. In principle, the weather ought to be entirely predictable. There is, for instance, a mathematical formula called the Navier–Stokes equation that determines completely the future evolution of a fluid such as the Earth's atmosphere. In practice, however, what the Navier–Stokes equation predicts depends enormously on the initial conditions. Plug into the equation two sets of temperatures at different locations around the world and the result, within a week, will be two entirely different weather systems.

American meteorologist and broadcaster Robert T. Ryan puts in a nutshell the challenge faced every day by weather forecasters: 'Imagine a rotating sphere that is 8,000 miles in diameter, with a bumpy surface, surrounded by a 25-mile-deep mixture of different gases whose concentrations vary both spatially and over time, and is heated, along with its surrounding gases, by a nuclear reactor 93 million miles away. Imagine also that this sphere is revolving around the nuclear reactor and that some locations are heated more during parts of the revolution. And imagine that this mixture of gases receives continually inputs from the surface below, generally calmly but sometimes through violent and highly localized injections. Then, imagine that after watching the gaseous mixture, you are expected to predict its state at one location on the sphere one, two, or more days into the future.'[14]

It is often said, in fact, that the weather is chaotic. This is a type of behaviour that shows infinite sensitivity to initial conditions. 'Does the flap of a butterfly's wings in Brazil set off a tornado in Texas?' asked Edward Lorenz, one of the pioneers of the mathematical theory of chaos.[15] The answer appears to be yes and no. Certainly, the atmosphere – particularly the circulation band at mid-latitudes – has the unpredictability of water boiling in a saucepan. However, it appears to hover somewhere between predictability and chaos. After all, if this were not the case, and Lorenz's butterfly effect held sway, weather forecasters would have no success at all. To many this is cold comfort. 'The trouble with weather forecasting', said the American financial analyst Patrick Young, 'is that it's right too often for us to ignore it and wrong too often for us to rely on it.'

The oceans

I have not mentioned the oceans. This is a big omission. The oceans are responsible for transporting about half of the heat from the equator to the poles, making them as important as the atmosphere.

In the North Atlantic, for instance, warm water from the Gulf of Mexico travels north past the west coast of Europe, boosting the region's temperature significantly above that of other landmasses at comparable latitudes, such as Canada. Near the pole, some of the water freezes into sea ice, in the process of which salt is driven out, making the sea water saltier. Since salt is relatively heavy, the water sinks to the bottom of the ocean. There, it flows along the sea floor back to the Gulf of Mexico. The result is a conveyor belt in the ocean reminiscent of the Hadley cell conveyor in the atmosphere, with warm water flowing north, cooling and sinking, then returning south.

But the oceans do more than transport heat from the equator to the poles. They also *store* heat, which they later slowly release. This evens out variations in the temperature between, for instance, summer and winter.

The seasons arise because the Earth does not spin with its equator always pointing towards the Sun. It spins tilted at 23.5° to the vertical. This means that, at one point in the Earth's orbit, the northern hemisphere is tipped towards the Sun, creating summer (*winter* in the southern hemisphere) and, six months later, tilted away from the Sun, creating winter (*summer* in the southern hemisphere). The Earth's orbit is not circular but elliptical, and summer in the south coincides with the time when the Earth is at its closest to the Sun.[16]

Because the equator does not always point towards the Sun, the hottest point on the surface of the Earth is not always the equator. In fact, the subsolar point migrates north and south with the seasons, and, with it, migrate the whole system of three circulation bands in each hemisphere.

The oceans play an important role in all of this because they store heat – a lot more than the atmosphere. Warmed up in summer, they then gradually release their heat during winter. This means that the coldest part of the winter in any hemisphere comes not at winter solstice – when that hemisphere is pointed away from the Sun – but several months later. Just as the atmosphere evens out extremes of temperature between *day and night*, the oceans even out extremes of temperature between *summer and winter*.

Climate

Climate, in contrast to weather – the day-to-day variation in atmospheric conditions – is defined as 'the average state of the atmosphere and oceans over longer periods of time than associated with weather'. Typically, this is of thirty years or more. 'Climate is what we expect,' said Mark Twain. 'Weather is what we get.'

One of the striking discoveries of science is that the climate of the Earth has not always been as it is. For instance, a whopping 90 per cent of the past 1 million years has been an ice age, a period of depressed global temperatures characterised by extensive ice sheets in the northern and southern hemispheres.

One of the triggers of ice ages is believed to be natural cycles in the Earth's orbit around the Sun caused by the gravitational tug of the Sun, Moon and other planets. Over a period of 100,000 years, for instance, the Earth's elliptical orbit becomes more stretched out than squashed up. Over a period of 42,000 years, the Earth's spin axis, currently tilted at 23.5° from the vertical, tips over further, then rears up closer to the vertical. And, over a period of 26,000 years, the Earth's spin axis changes its orientation in space, rotating through a full circle about the vertical.[17] These cycles, known as Milanković cycles, vary the amount of sunlight falling on the Earth's surface.

But, it is not only variations in the amount of sunlight intercepted by the Earth that are thought to cause ice ages. Intrinsic variations in the Sun may also play a role. The Sun is actually remarkably steady and alters its heat output by less than 1 per cent over the course of a solar cycle.[18] This is too little to have much effect on the Earth's climate. However, the small fluctuation in total heat output is accompanied by a variation of as much as 100 per cent in solar ultraviolet radiation.[19] Such high-energy light shatters high-altitude molecules such as ozone in the stratosphere, the layer above the weather, or troposphere. Since these molecules play an important role in transporting heat down through the atmosphere, the boost in solar ultraviolet can have an appreciable effect on the Earth's climate.

But it is not simply changes in the amount of sunlight falling on the Earth's surface that play a role in triggering ice ages. There are more down-to-earth things – literally – such as the movement of continents.[20] Once upon a time, for instance, South America was connected to Antarctica. Warm water flowed from the equator directly down the coast of South America, keeping Antarctica ice-free. About 33 million years ago, however, the two continents broke apart. With the opening up of the Drake Passage between South America and Antarctica, it was suddenly possible for water to circulate in a west-to-east direction between the Pacific and Atlantic oceans. With water now largely flowing from west to east, rather than from north to south, the flow of heat towards Antarctica was significantly reduced and Antarctica, as a consequence, froze.

A similar change to the ocean circulation could be triggered in the North Atlantic by human-induced global warming. Currently, warm water flows from the Gulf of Mexico up past the coast of western Europe. There it cools, sinks and returns. This conveyor belt of warm water keeps the coast of western Europe relatively warm. However, melting sea ice near the pole could disrupt the flow of heat from the equator to the pole. This is because sea ice, when it initially forms, expels salt. The melting of sea ice near the pole will therefore make the water less salty and, crucially, *less heavy*, so that it no longer sinks (melting fresh-water ice from Greenland will do the same). The result could be that the North Atlantic conveyor will to some extent shut down, plunging the temperature off the coast of Europe to a level more typical of its latitude – that is, more like Winnipeg in Canada. The Earth will, of course, still have to transport heat from the equator to the poles. But air currents and east–west flows, through the unfrozen Arctic sea, might take over that role, much as they did 33 million years ago after the split of South America from Antarctica.

Recent ice ages, however, are nothing compared with ancient ice ages. The world is believed to have gone through two periods when ice stretched in an unbroken sheet all the way from the poles to the equator. These episodes, known as Snowball Earths, occurred about 650 million years ago and 2.2 billion years ago, respectively. The causes are disputed. But a plausible explanation of the first episode is that it was caused by blue-green algae suddenly evolving the ability to split water molecules and release oxygen in photosynthesis. This happened about 2.3 billion years ago. The oxygen from such cyanobacteria destroyed methane – an abundant greenhouse gas in the atmosphere – which had been keeping the planet warm.

A planet covered entirely in ice reflects sunlight back into space. For this reason, Earth is likely to have remained locked in each of its Snowball states for millions of years. What brought each super-cold spell to an end was probably volcanic eruptions, which pumped more and more carbon dioxide back into the atmosphere until, finally, its warming effect was enough to thaw out the Earth.

Greenhouse warming

Carbon dioxide is of course the gas that is produced by the burning of fossil fuels such as oil and coal and whose concentration in the atmosphere has been increasing since the beginning of the industrial age. Over precisely the same period the global temperature has been steadily rising – exactly what would be expected since carbon dioxide is known to trap heat in the atmosphere.

It works this way. Carbon dioxide – and the rest of the gases that compose the atmosphere – are transparent to visible light from the Sun (if they were not, we would not be able to see the Sun). Sunlight therefore passes through the air unhindered and heats the ground. The ground, in turn, heats the air, which is why the temperature is highest near the ground and steadily decreases with altitude all the way to the top of the troposphere, the domain of weather.

To be precise, the ground glows with heat radiation typical of a body at about 20 °C. Crucially, such far infrared is absorbed by carbon dioxide in the atmosphere. In other words, the Earth's heat is prevented from escaping into space and is instead trapped in the atmosphere. This is not quite what happens in a greenhouse, where glass is transparent to sunlight but provides a *physical barrier* to the escape of rising, or convecting, warm air. Despite this, however, carbon dioxide is widely known as a greenhouse gas.

Actually, by far the most important greenhouse gas in the atmosphere is water vapour. This is responsible for about 75 per cent of the warming effect of the atmosphere compared with only 20 per cent for carbon dioxide. We should on the whole be grateful for greenhouses gases since, without them, the average temperature of the Earth would be a super-chilly -18 °C.

However, if humans continue adding more and more carbon dioxide to the atmosphere, the global temperature will continue to rise. 'Geological change usually takes thousands of years to happen but we are seeing the climate changing not just in our lifetimes but also year by year,' warned the English chemist James Lovelock.

The Greenland ice sheet and Antarctic ice sheet are already melting. But the melting will accelerate, significantly raising the sea level globally and inundating low-lying coastal areas. The circulation of the ocean and atmosphere will change unpredictably, with worrying implications for the Earth's 7 billion people. Nobody knows where it will all end. However, nature has conveniently shown us one possibility: Venus.

Being about two-thirds of the Earth's distance from the Sun, Venus lost its water early on in its history. Basically, the extra heat from the Sun caused its primordial oceans to begin evaporating away. Water vapour, being a potent greenhouse gas, warmed the planet more, which evaporated more of the oceans, which warmed it even more, and so on. This runaway greenhouse effect, first proposed by Carl Sagan and William Kellogg in 1961, eventually boiled away Venus's oceans entirely. We see no sign of them today because, at the top of the atmosphere, high-energy ultraviolet from the Sun split water molecules into their constituent hydrogen and oxygen atoms, which then wafted away from the planet on the wind from the sun. Ultimately, Venus lost its oceans to space.

On Earth, carbon dioxide from volcanoes is washed out of the atmosphere by rain. But this could not happen on waterless

Venus. Instead, the level of carbon dioxide in the atmosphere rose and rose. Today, the planet has about 92 Earth-atmospheres-worth of carbon dioxide. Not only does this create a crushing pressure on the surface – equivalent to the pressure almost a kilometre down in the Earth's oceans – but the warming effect of the greenhouse gas creates a temperature hot enough to melt lead. The whole planet is shrouded in impenetrable sulphuric-acid clouds, made from sulphur dioxide vomited from volcanoes. Venus, in short, is hell.

Since the Earth is further from the Sun than Venus, it is not clear whether our warming of the planet will eventually trigger the catastrophe of a runaway greenhouse. But, whether we are responsible or not, one thing is sure: one day it will happen *naturally*.

The reason is that the Sun is slowly growing hotter as it burns through its hydrogen fuel.[21] In fact, it is now about 30 per cent brighter than it was at its birth, 4.55 billion years ago.[22] In the future, as the Sun continues to get more luminous, more and more water from the oceans will turn into water vapour, which will trap more heat in the atmosphere, which will turn more of the oceans into water vapour, and so on. On the sweltering planet, carbon dioxide, locked up in carbonate rocks such as chalk cliffs, will begin leaking into the atmosphere, trapping more heat, which will create more heating, which will drive more carbon dioxide into the atmosphere. Eventually, by about AD 1 billion, the oceans will have boiled away entirely into space and the atmosphere will be made mostly of carbon dioxide. Coincidentally, the Earth has pretty much the same amount of carbon dioxide locked up in carbonate rocks as Venus currently has in its super-dense atmosphere. So, when it all floods out into the atmosphere, the Earth will be *almost exactly like Venus*.

But this will not be the end of the Earth's ordeal. In 5 billion years' time, the Sun will run out of hydrogen fuel in its core. It will swell into a monstrous super-luminous red giant, pumping out 10,000 times as much heat as it does today. If this bloated star does not completely swallow our planet – and it will definitely envelop the close-in worlds of Mercury and Venus – it will certainly reduce the Earth to a burnt and blackened lump of slag.[23]

Long before that time, however, our descendants – should any still survive – will have to leave the Solar System and find another planet to live on. 'Earth is the cradle of humanity,' said Siberian rocket pioneer Konstantin Tsiolkovsky. 'But mankind cannot stay in the cradle for ever.'

PART FOUR: Deep workings

14:

WE ARE ALL STEAM ENGINES

Thermodynamics

Not knowing the second law of thermodynamics
is equivalent to never having read a work by
Shakespeare.
c. p. snow, 'The Two Cultures', 1959

All our actions, from digestion to artistic creation,
are at heart captured by the essence of the operation
of a steam engine.
PETER ATKINS, *Four Laws that Drive the Universe*

How much energy does the Earth trap from the Sun? The answer is zero. Think about it. On a hot day, out in the Sun, you sweat. By this means, you shed heat exactly as fast as your body absorbs it. If you did not, you would get ever hotter until eventually you keeled over from heat exhaustion. Similarly, the Earth radiates heat back into space at the rate it receives it from the Sun. If it did not, it would get hotter and hotter until its rocks turned to the consistency of honey.[1]

But, if the Earth is not gaining any net energy from the Sun, *what* is it gaining? After all, something is powering all activity, including biological activity, on Earth. The clue is to look beyond the *amount* of energy arriving from the Sun to the *quality* of that energy. The heat radiated by the Earth back into space turns out to be of a *poorer quality* than the heat that is intercepted by the Earth from the Sun. The Earth saps something from it. But what?

To answer this, it is necessary to know that the Earth is like a steam engine. In fact, as English chemist Peter Atkins says, 'We are all steam engines.'[2] Don't be perturbed by this. A steam engine is, in essence, a very simple device. Basically, it consists of a container at a high temperature filled with steam. The steam pushes a movable wall in the container – a piston – against the outside air pressure. Having done this, the steam ends up condensed as water at a low temperature – the temperature of the surrounding air. That's it.

Focus for a moment on the piston. When something moves against a force it is said to do work. That is what the piston does as it moves against the outside air pressure. Work is what everything on Earth is doing today. Your muscles do work each time you lift your foot against the force of gravity. The electrons in the current flowing through your computer do work as they move against the resistive force of atoms blocking their path. Without work there would be no activity. Everything in the world would just sit there, inert, inactive, for all eternity.

In the case of a steam engine, work is done by heat energy, which starts out at a high temperature and ends up at a low temperature. And it is exactly the same for the Earth. Work is done by heat energy that starts at a high temperature – the 5,500 °C characteristic of the Sun – and ends up at a low temperature – the 20 °C typical of the Earth's surface. However, instead of driving a mere piston, this energy drives everything from the swirling of hurricanes to the swimming of fish to the biochemical reactions that keep your body at a liveable 37 °C.

Clearly, then, it is the *temperature* that makes the heat energy radiated by the Earth qualitatively different from the heat energy intercepted by the Earth.[3] But why is a change in temperature associated with work? To answer this it is necessary to understand what heat and work *are*.

Heat is disordered motion. If you could see the molecules in steam you would see them flying about randomly like a swarm of angry bees. If you could see the atoms in a white-hot bar of iron, you would see them jiggling randomly about their fixed positions. Work, by contrast, is ordered motion. If a piston moves, or a muscle in your arm contracts, a large number of atoms move as one, in lockstep.

So here is what happens when steam in a container does work by pushing a piston. Molecules of steam drum on the piston like countless raindrops on a tin roof.[4] Although each molecule has only a tiny pushing effect, together they have enough oomph to drive the bulk of the piston.

Now, temperature is a measure of the average speed of a body's microscopic constituents such as atoms; whereas the atoms of a hot body are moving quickly, those of a cold one are moving more sluggishly. And each molecule of steam, in imparting its tiny pushing force to the piston, loses some of its energy of motion, some of its speed.[5] So the price of doing work on the piston is a drop in the average speed of the molecules – in other words, a reduction in the steam's temperature.

But there is a subtlety here. And it will bring us back to the case of the Earth and the Sun – and arguably one of the most profound insights in the whole of science: the second law of thermodynamics.

How much heat can be converted into *useful* work for the piston? Naively, you might think, all of it. After all, one of the basic laws of physics – actually, the first law of thermodynamics – states that energy cannot be created or destroyed, only converted from one form into another. For instance, electrical energy can be converted into light energy and heat energy in a light bulb; chemical energy can be converted into the energy of motion of muscles in your body, and so on. However, the law of conservation of energy – its more common name – tells us only what is possible in principle, not what is possible *in practice*.[6]

The problem in harnessing steam to drive a piston is that it uses *random* microscopic motion to drive *ordered* bulk motion. While some of the steam atoms are flying in exactly the direction of motion of the piston, many are not, and some are even flying at right-angles to the piston's direction, making them useless at pushing it. Clearly, not all the energy of motion of the molecules of the steam is *usable*, not all of it can be converted into the energy of motion of the piston. A steam engine, consequently, can never be 100 per cent efficient.[7]

This is in fact a statement of the second law of thermodynamics, as formulated in the nineteenth century by the British physicist Lord Kelvin: 'Heat cannot be turned into work with 100 per cent efficiency.' This is not a very exciting or insightful version of the law – certainly not one that hints at its all-conquering power and its profound implications for life, the Universe and everything. That requires understanding the concept of entropy.

Entropy, in broad-brush terms, is the degree of microscopic disorder of a system such as a container filled with steam. When an amount of heat, Q, is added to the system at a temperature, T, its entropy, S, increases by Q/T. If this seems baffling, there is some common sense behind it.

Temperature is a measure of how vigorously atoms are moving about. Take a low-temperature body. Adding heat to it is like sneezing in a quiet environment such as a library. It has a big effect – that is, there is a large increase in the disorder, or entropy. Contrast this with a high-temperature body. Adding exactly the same amount of heat as before is like sneezing in a bustling shopping street. It has little effect – that is, there is only a small increase in the disorder, or entropy.

In a steam engine, the steam is initially at a high temperature. The energy that leaves it to drive the piston therefore reduces the entropy of the steam – but only by a small amount (remember the sneeze in the bustling street). But the energy ends up in the surrounding area at a lower temperature. This boosts the entropy to the surroundings by a large amount (remember the sneeze in the library).[8] In other words, the result of doing work on the piston is a net *increase* in the entropy of the system and its surroundings.

And this *always* happens. When work is done, the entropy of the Universe always increases. In fact, this is the definitive statement of the second law of thermodynamics, arguably the most far-reaching of all laws of physics. 'The law that entropy always increases holds, I think, the supreme position among the laws of Nature,' said British physicist Arthur Eddington. 'If your theory is found to be against the second law of thermodynamics, I can give you no hope; there is nothing for it but to collapse in deepest humiliation.'[9]

In the case of the Earth, heat energy with a characteristic temperature of 5,778 Kelvin[10] is absorbed from the Sun, and heat energy with a temperature of about 300 Kelvin is re-radiated into space.[11] This corresponds to an enormous net increase in entropy.[12] It is the price the Universe pays for the bewildering multitude of work processes driven on the Earth.

Contrary to popular belief, the Earth does not have an energy crisis – it uses essentially no net energy from the Sun. What the Earth actually has is an *entropy crisis*. Once heat has done work, it becomes heat at a lower temperature, which means it is more disordered, lower-quality energy, with a more limited ability to do any more work. Recall, after all, that more work can be done by a steam engine only if the heat is expelled at an even lower temperature. However, practically, there is a rock-bottom limit set by the temperature of the surrounding air. Once heat is at this temperature, it is impossible to expel it at a lower temperature and so no further work can be done.

How life bucks the trend of increasing entropy

But, if entropy – disorder – always increases, how come we appear to live in a world of order? In particular, how do living things, which are structured and as far from disordered as it is possible to imagine, buck the trend of increasing entropy?

The answer is that the second law insists only that entropy increases *overall*. So, in the creation of a cell, heat is generated by all the chemical reactions needed to assemble the cell membrane and internal cellular machinery. That heat boosts the entropy of the surroundings far more than the assemblage of the cell reduces it. Life exports disorder to the Universe.

And, just as all processes on Earth are ultimately driven by the temperature difference between the Sun and the Earth, *all processes in the Universe* are ultimately driven by the temperature difference between the stars, which are hot, and empty space, which is cold.[13] Remember that next time you are out on a crystal-clear night and gaze upwards at the stars.

But this still does not explain how life bucks the trend of ever-growing disorder. After all, whenever work happens spontaneously – for example, when a slate falls off a roof, smashing on the ground, creating heat energy and sound energy – entropy always increases. The answer is that life is clever. Think of a big weight hoisted on a pulley to a great height (the analogue of a high temperature). If the weight falls, it can do work – perhaps driving the hands of a grandfather clock. At the same time, heat is inevitably generated – through friction of the rope on the pulley, friction in the clock mechanism, and so on – boosting entropy. But, say, things have been set up so that, as the weight falls, it hoists into the air a smaller weight. That creates a little order, reducing entropy slightly. Later, if the smaller weight falls, it can do work. But say things have also been set up so that, as the small weight falls, it hoists into the air an even smaller weight, creating a little more order, reducing entropy a little more.

This is how life feeds off heat energy, degrading it to even lower-quality heat energy while at the same time creating more order. Of course, it does not use weights and pulleys – it employs a whole series of other tricks – but the principle is the same. Plants, for instance, absorb sunlight and use it to create energy-rich chemicals, the equivalent of raising a small weight that can be used later to do work. In a bewildering cascade of other chemical processes, life wrings every drop of work it can from solar energy, finally discarding it as low-quality heat, the ultimate slag of the Universe.

Life creates order at the expense of a lot more disorder, which it exports to its surroundings. The Earth's biosphere is an island of organisation in a cosmic sea of chaos. 'Life is nature's solution to the problem of preserving information[14] despite the second law of thermodynamics,' says Howard Resnikoff.[15]

Heat Death

Of course, when looked at globally, no process in the Universe defies the trend of remorselessly increasing disorder. In the nineteenth century, when this was first recognised by physicists, it deeply depressed them. After all, if entropy is continually increasing, it stands to reason that, sooner or later, the Universe will reach a state of maximum entropy. At this point, all heat in the Universe will be reduced to its lowest-grade state. There will be no temperature differences to drive any activity. In this state of cosmic ennui, which the nineteenth-century German physicist Rudolf Clausius called Heat Death, all cosmic machinery will come to a juddering halt. The Universe, in the words of the poet T. S. Eliot, will end 'not with a bang but with a whimper'.[16]

As far as we are aware, there is no way the Universe can escape such a miserable fate. An interesting question therefore is: how close is the Universe to Heat Death? The answer is: much closer than you might imagine. Although it might seem that countless stars pumping out random starlight across the length and breadth of the Universe account for its disorder, this is an illusion. Most of the disorder in the Universe is in fact tied up in the afterglow of the fireball of the big bang. Incredibly, 13.8 billion years after the beginning of time, this cosmic background radiation still permeates every pore of the Universe. Greatly cooled by the expansion of the Universe over the past 13.8 billion years, it now appears as far infrared, a type of light invisible to the naked eye. Whereas starlight accounts for a mere 0.1 per cent of all the photons, or particles of light, in the Universe, the cosmic background radiation accounts for a whopping 99.9 per cent.

The key thing to know is that the fireball radiation broke free of matter – *was created*, in essence – a mere 379,000 years after the big bang. Consequently, the Universe, for most of its existence, has, in relative terms, been close to Heat Death. Despite this, there is still plenty of scope for cosmic entropy to increase. There are a few tens of trillions years to go until the stars have all gone out and the Universe is plunged into a night without an end.

Take heart from the German physicist Arnold Sommerfeld. Discussing thermodynamics, he wrote, 'The first time you go through it, you don't understand it at all. The second time you go through it, you think you understand it, except for one or two small points. The third time you go through it, you know you don't understand it, but by that time you are so used to it, so it doesn't bother you any more!'

15:

MAGIC WITHOUT MAGIC

Quantum Theory

God does not play dice with the Universe.

ALBERT EINSTEIN

Stop telling God what to do with his dice.

NIELS BOHR

Quantum theory is our very best description of the microscopic world of atoms – the building blocks of ordinary matter – and their constituents. It is a fantastically successful theory. Not only has it given us lasers and computers and nuclear reactors but it has provided an explanation of how the Sun shines and why the ground beneath our feet is solid.

But, in addition to being a fantastic recipe for making things and understanding things, quantum theory provides a unique window on the weird, counter-intuitive, *Alice-in-Wonderland* world that underpins everyday reality. It is a place where a single atom can be in two places at once; where things happen for absolutely no reason at all, and where two atoms can influence each other *instantaneously* even if they are on opposite sides of the Universe.

How did quantum theory come about?

Quantum theory was born out of a conflict between two great theories of physics – the theory of matter and the theory of light. The theory of matter holds that, ultimately, everything is made of tiny indivisible grains, or atoms.[1] The theory of light says that light is a wave, spreading outwards from its source like a ripple on a pond.

Both theories are very successful. For instance, the theory of atoms explains the behaviour of gases such as steam. If a gas is squeezed into half its volume, it pushes back with twice the force, or pressure, an observation encapsulated in Boyle's law. This can be explained if the pressure is caused by countless tiny atoms drumming on the walls of the container like rain on a tin roof. When the volume is halved, the atoms have only half as far to fly between striking and restriking the walls and so drum twice as often on the walls, doubling the pressure.

The theory of light is also very successful. However, the phenomena it explains generally involve light waves that overlap each other and reinforce or cancel each other out. And, since the distance between successive crests of a light wave is far less than the width of a human hair, such interference or diffraction phenomena are hard to spot and take scientific ingenuity to make visible to the naked eye.

The clash between the theory of light, which says light is a wave, and the theory of matter, which maintains matter is made of atoms, occurs not surprisingly in the place where *light meets matter*. Specifically, when an atom spits out light – for instance, in a light bulb – or when an atom gobbles up light – for example, in your eye.

The problem is not hard to appreciate. A light wave is fundamentally a spread-out thing whereas an atom is fundamentally a localised thing – it would take 10 million laid side by side to span the full stop at the end of this sentence. In fact, a wave of visible light is about 5,000 times bigger than an atom. Imagine you have a matchbox and you open it and out drives a 40-tonne lorry. Or, alternatively, a 40-tonne lorry approaches, you open a matchbox, and the lorry slips inside. That's the way it is when light meets matter. Somehow an atom must swallow or cough out something 5,000 times bigger than itself.

Logically, the only way something can be emitted and absorbed by something as small and localised as an atom is if it too is small and localised. 'Nothing fits inside a snake like another snake,' observed TV survival expert Ray Mears. The trouble is there are countless experiments that show unequivocally that light is indeed a spread-out wave.

Resolving the paradox was mental torture for the physicists of the 1920s. 'I remember discussions . . . which went through many hours until very late at night and ended almost in despair,' wrote the German physicist Werner Heisenberg. 'And when, at the end of the discussion, I went alone for a walk in the neigh-bouring park I repeated to myself again and again the question: Can nature possibly be so absurd as it seemed to us in these atomic experiments?'[2]

In the end physicists were forced to accept something scarcely believable: that light is both a spread-out wave *and* a localised par-ticle. Or, rather, light is neither a wave nor a particle. It is *something else* for which we have no word in our language and nothing with which to compare it in the everyday world. It is as fundamentally ungraspable as the colour blue is to a person blind from birth. 'We must content ourselves with two incomplete analogies – the wave picture and the corpuscular picture,' said Heisenberg.[3]

In retrospect, perhaps physicists should not have been sur-prised to find the submicroscopic world *weird*. Why should the world of the atom – which is 10 billion times smaller than a human being – contain objects that behave in any way like those in the everyday world? Why should they dance to the same tune, the same laws of physics?

Light is an ungraspable thing and all we can ever do is observe the facets of it. When light is absorbed or spat out by an atom, we see its particle-like face, known as a photon. When light bends around a corner, we see its wave-like face.[4] 'On Mondays, Wednesdays and Fridays, we teach the wave theory and on Tuesday, Thursdays and Saturdays the particle theory,' joked the English physicist William Bragg in 1921.

But, it turns out, things are much worse than this. In 1923, the French physicist Louis de Broglie, writing in his doctoral thesis, proposed that not only can light waves behave as localised particles, particles such as electrons can behave as spread-out waves. According to de Broglie, all the microscopic building blocks of matter have two faces. All share a peculiar wave–particle duality. In fact, if there is one thing you need to know in order to understand quantum theory – one thing from which pretty much everything else logically follows – it is this: *Waves can behave as particles and particles can behave as waves.*

Waves as particles imply unpredictability

Take the first half of the sentence first: waves can behave as particles. Imagine you are looking though a window at the street outside. Maybe you see a car going past, a woman walking her dog past a tree. If you look closely, however, you will also see a faint reflection of your own face staring out. This is because glass is not perfectly transmitting. Most of the light – say, 95 per cent – goes right through but the remainder – 5 per cent – is reflected back. The question is: how is this possible if light behaves as particles – a stream of identical photons like so many miniature machine-gun bullets?

Surely, if the photons are all identical, they should all be affected *identically* by the window pane? Either they should *all* be transmitted or *all* reflected. There appears to be no way to explain how *some* can be transmitted and *some* reflected. Unless – and here physicists were forced to accept a diminished, cut-down version of what it means to be *identical* in the microscopic world – photons have an identical *chance* of being transmitted, an identical *chance* of being reflected.

But this, as Einstein first realised, is catastrophic for physics. Physics is a recipe for predicting the future with 100 per cent certainty. The Moon is over here today and Newton's theory of gravity predicts where it will be tomorrow with *absolute confidence*. But, if photons merely have a particular chance of being transmitted, then it is impossible to predict what an individual photon will do when its strikes the window pane. Whether it goes through or bounces back is entirely down to chance.

And we are not talking about the kind of chance with which we are familiar in the everyday world. We may think a roulette ball ends up where it ends up by chance. But, actually, if we knew the initial motion of the ball, the friction between the wheel and the ball, the play of air currents in the casino, and so on, Newton's laws would predict *exactly* where the ball would end up. The fact we cannot do this is merely down to not being able to measure all these things accurately enough and do the required calculation to enough decimal places. Though we could do it in principle, we could not do it in practice. However, when we come to a photon impinging on a window, what it does – whether it is transmitted or reflected – is not even predictable *in principle*. Quantum unpredictability is *truly* something new under the Sun.

And it turns out that it is not just photons that are fundamentally unpredictable. So too are *all* the denizens of the submicroscopic world, from neutrons to neutrinos, electrons to atoms.

Einstein was so appalled by this that he famously declared: 'God does not play dice with the Universe.' But Einstein was wrong.[5]

An obvious question arises: if the Universe at its fundamental level is unpredictable, how come we know the Sun will rise tomorrow, that a ball will go roughly where we throw it? The answer is that what nature takes away with one hand it grudgingly gives back with the other. Yes, the Universe is unpredictable. But, crucially, the *unpredictability is predictable*. In fact, this is what quantum theory *is*: a recipe for predicting the unpredictable – the probability of this event, the probability of that event. And this, it turns out, is enough to create the largely predictable world we find ourselves in.

The fact that, ultimately, things happen randomly, for no reason at all – the consequence of waves behaving as particles – is arguably the most shocking discovery in the history of science. But, recall, there is a second half to that crucial sentence: particles can behave as waves. The consequence of this turns out to be equally stunning.

Particles as waves imply superpositions

Clearly, if particles can behave as waves, they can do *all* the things waves can. And one thing in particular waves can do is mundane in the everyday world but has truly earth-shattering consequences in quantum world.

Imagine there is a storm out at sea and huge waves are rolling in to a beach. Imagine that the next day the storm has passed and the surface of the sea is ruffled into small ripples by a gentle breeze. Now, anyone who has watched the sea knows that it is possible to have a wave that is both big and rolling *and* that also has small ripples on its surface. This is a general property of all waves. If two waves can occur individually, it is always possible to have a combination, or superposition of the two.

Now consider a quantum wave associated with, say, an atom. Actually, this is a slightly peculiar wave because it is a mathematical thing. Nevertheless, it can be imagined extending throughout space. The important thing is that where it is big there is a high probability of finding the atom and where it is small there is a low probability.[6]

So far, nothing untoward.

Now imagine two quantum waves. One is a quantum wave for an oxygen atom that is highly peaked 10 metres to your left, so there is a very high probability of finding it there. And the other is a quantum wave for the same oxygen atom that is highly peaked 10 metres to your right, so there is a very high probability of finding it there. But, recall, it is a general property of waves that, if two waves are possible, so too is a superposition of the two. But, in this case, such a combination will correspond to an oxygen atom that is simultaneously 10 metres to your left and 10 metres to your right – in other words, in *two places at once*. That is the equivalent of you being in London and New York at the same time.

Actually, nature is set up in such a way that it is impossible to observe something being in two places at once. That is because, if we try to locate something, we are implicitly looking for its particle-like property, which precludes seeing a wave-like property such as superposition. So who cares? Well, although it is impossible to observe something being in two places at once, it is nevertheless possible to observe the *consequences* of something being in two places at once. The wave phenomenon that makes this possible and spawns all kinds of quantum weirdness is called interference.

Interference

If you have ever seen raindrops falling in a pond, you will have seen concentric ripples spreading out from each impact and over-lapping with each other. Where the crests of two waves coincide, they reinforce each other, making a bigger wave; where the crest of one wave coincides with the trough of another, they cancel each other, creating dead calm. This is interference. Now, im-agine inserting a piece of card in the region of overlap of the waves spreading from two raindrops. There will be places on the card where big waves strike and there will be places on the card where no waves hit.

Actually, this experiment was done with light by the English physician and polymath Thomas Young in 1801. With consider-able ingenuity, he managed to engineer a situation in which there was an overlap between the light spreading from two point sources of illumination. When he inserted a screen in the over-lapping region, there appeared a pattern of alternating light and dark stripes, not unlike a modern-day barcode. It was undeniable proof that light exhibits the characteristic wave phenomenon of interference. Young had proved that light ripples through space like an undulation on the surface of a pond. Nobody had noticed it before because the waves of light are simply far too small to be seen with the naked eye.

Because of interference, the fact that a quantum object such as an electron can be in two places at once *has consequences*. Here is an example. Imagine two bowling balls that are rolled together so they collide and ricochet off each other. If this happens over and over again, the balls will be seen to scatter in a range of different directions. Imagine a giant clock face. The balls will go to every number on the clock face.

Now imagine two quantum particles – say, electrons – which collide and scatter in a similar manner. If this happens thousands upon thousands of times, the electrons will also scatter in a range of different directions. But something very odd will soon become apparent. Some directions will be favoured by the electrons. And others will be studiously avoided. In other words, there will be numbers on the clock face where the electrons *never go*.

The explanation is that there are directions in which the electron waves reinforce each other and directions in which they cancel each other out. The latter are the directions in which no electrons are seen.

This interference phenomenon was demonstrated in 1927 by Clinton Davisson and Lester Germer in the US and by George Thomson in Scotland. The physicists bounced electrons off the flat surface of a crystal and noticed that there were directions in which the electrons *never* bounced. The crystal consisted of layers of atoms like a loaf of sliced bread stood on end. Some electrons bounced off the top layer; some off the layer below; some off the layer below that, and so on. And the quantum waves of all these electrons *interfered* with each other. Only in the directions where all the waves reinforced did the experimenters observe electrons.

By showing that electrons can interfere with each other, proving that electrons are indeed waves, Davisson, Germer and Thomson won the Nobel Prize for Physics. The irony is that, while Thomson received the Nobel for showing that the electron is not a particle, his father, 'J. J.' Thomson, had received the Nobel for showing that *it is*.[7] There can be no better illustration of the paradox at the heart of quantum theory.

What all this shows is that, even though it is not possible to see a single quantum particle go in several directions at once, interference means there are consequences. The quantum waves corresponding to the electron going in all possible directions interfere with each other, reinforcing in some directions and cancelling each other out in other directions. That is why there are directions that electrons never go. That is why there is quantum weirdness.[8]

Currently, there is a race on in the world to exploit superpositions – the ability of atoms and the rest to do many things at once – to do *many calculations at once*. People are trying to build a quantum computer, which promises to outperform massively even the most powerful conventional computer with certain types of calculations. The reason for saying 'certain types of calculations' is that they must have a single answer. Recall that it is impossible to observe a quantum particle doing many things at once, merely the *consequence* of it doing many things at once. Similarly, it is impossible to access all the countless individual strands of a quantum computation, only the consequence – that is the single answer made from all the threads woven together.

Why is the everyday world not quantum?

Building a quantum computer is extremely difficult because, if the quantum building blocks of such a computer interact with their surroundings in any way, the multi-tasking power of the computer is irrevocably lost. So a quantum computer must be totally isolated in a vacuum chamber so that no air atoms strike it or photons of light. And this is hard.

It is not that the ability of the quantum particles to do many things at once is fragile. It is simply that it is very difficult for a large number of atoms – such as air atoms – to maintain a super-position. If the quantum particle impresses its superposed state on a lot of atoms, the impression is quickly lost, a bit like one voice being drowned out in a crowd of chanting football supporters.

This explains why atoms display quantum weirdness but, when large numbers of atoms come together to make everyday objects, those objects do not display quantum weirdness. For instance, you never see a table in two places at once or someone walking through two doors at the same time. We never see quantum behaviour in the everyday world because we never see individual atoms or photons. We see only large numbers. You do not observe the world; you observe yourself. In other words, your brain never observes a photon; it observes the amplified effect of that photon impressed on hundreds of thousands of atoms in your retina. And that impression loses the quantumness of the original photon. This is why, bizarrely, we live in a quantum world that *does not look quantum*.

Quantum weirdness

Much quantum weirdness is a consequence of superpositions and interference. But there are other quantum ingredients also. And, when they are combined in different permutations, they spawn all sorts of novel and surprising behaviours. 'Magic without magic', as it has been called. Take quantum spin. This is a property, like quantum unpredictability, that has no analogue in everyday life. Basically, a quantum particle behaves as if it is spinning like a tiny top, even though it is not. Physicists say it has *intrinsic* spin, or angular momentum.

An electron has the smallest possible quantity of spin, which for historical reasons is called spin ½[9] rather than spin 1, which would be the sensible thing to call it.[10] Now a spinning charge acts like a tiny magnet.[11] This means that it acts like a compass needle when in a magnetic field, aligning itself either pointing along the field (up) or against it (down). If there are two electrons, one possibility is that electron 1 is spin up and electron 2 spin down; another possibility is that 1 is down and 2 is up. Now here is the important thing. In the quantum world superpositions are possible. So the two electrons can be up–down and down–up *at the same time*. A bit like you being simultaneously dead *and* alive.

So much for ingredient 1 – superposition. Ingredient 2 is the law of conservation of angular momentum. Basically, this says that the total spin of the two electrons can never change. Since the two electrons in the above example begin pointing in opposite directions, they must *always* point in opposite directions. Ingredient 3 is simply quantum unpredictability. If we observe an electron, whether it turns out to be spin up or down is fundamentally unpredictable like a quantum coin toss. There is a 50 per cent chance of it being up and a 50 per cent chance of it being down.

If all this is getting complicated, here is the situation where the three ingredients come together to create something extraordinary. We start with the pair of electrons that is in a superposition of up–down and down–up, and send one a long way away. When we have done this, we look at the stay-at-home electron. Perhaps we find that its spin is up. If so, instantaneously, its partner, far away, must flip down since the two spins must *always point in opposite directions*. Perhaps we find that its spin is down. Instantaneously, its partner must flip up.

What is so surprising about this is that, even if the far-away electron was on the other side of the Universe, it would still have to react *instantaneously* to its partner being found to be up or down. To Einstein this 'spooky' action at a distance, apparently in violation of the cosmic speed limit set by light, was so ridiculous that it *proved* quantum theory was flawed.[12] But, not for the first time, Einstein was wrong. Non-locality was triumphantly demonstrated in a laboratory in Paris by French physicist Alain Aspect in 1982.

It is worth pointing out that separating the two electrons is not like separating a pair of gloves. Clearly, if the stay-at-home glove is found to be the left one, the distant one will be a right one. That is because one glove was a left-hand one and the other a right-hand one *at the outset*. But the two electrons were neither up nor down at the outset. Their state was undetermined. The stay-at-home electron assumed its state *only when it was observed*. And that state was *random*. This is why non-locality does not violate Einstein's special theory of relativity. If up and down were like the dots and dashes of Morse code, all that could ever be sent would be a random sequence of dots and dashes because the state of the stay-at-home electron and its far-away cousin would always be selected randomly. The dots and dashes could not be controlled. Special relativity, it turns out, limits only the speed of a *meaningful signal*. Nature does not care about unusable garbage. It is welcome to fly about the Universe at any speed it likes. Nobody knows how this happens. Non-locality is arguably the deepest mystery of quantum theory.

Atoms: Why they exist at all

But quantum theory's greatest achievement is in explaining atoms. 'Atoms are completely impossible from the classical point of view,' said Richard Feynman. According to Maxwell's theory of electromagnetism, an accelerated charge – one that changes its speed or direction or both – radiates into space electromagnetic waves.[13] An electron orbiting an atomic nucleus is continually changing its direction. It should therefore broadcast like a tiny TV transmitter and rapidly lose energy. Calculations in fact show it should spiral into the nucleus in less than a hundred-millionth of a second. Atoms, as Feynman observed, have no right to exist.

Quantum theory comes to the rescue because quantum theory recognises that an electron has a wave nature. And it turns out that the smaller the mass of a particle the bigger its quantum wave.[14] Because you are so big, your wavelength is ridiculously tiny. This is why you exhibit no obvious wave behaviour. This is why you do not bend around corners or pass by on both sides of a lamp post. But the electron is the smallest particle in nature. It is precisely because it has the biggest quantum wave that it exhibits so much quantum weirdness. And its wave nature explains the existence of atoms. A wave is a fundamentally spread-out thing. It simply cannot be squashed into a nucleus.[15] So atoms do not shrink down to oblivion in a hundred-millionth of a second. Instead, they can exist essentially for ever.

In fact, the electron wave needs so much room that it explains another puzzling feature of atoms: why an electron orbits so far from its nucleus. An atom is 99.9999999999999 per cent nothingness.[16] You are 99.9999999999999 per cent nothingness.[17]

Atoms are empty – or so big compared with their nuclei – simply because an electron wave needs lots of elbow room.

But electron waves have a lot more to tell us about atoms. In fact, they explain *everything* about atoms.

Atoms: Why they come in different types

There is not just one kind of electron wave that can exist inside an atom. There are many. A more wiggly, more violent wave has more energy than a more sluggish one. It therefore corresponds to an electron that is capable of defying the pull of the nucleus and orbiting further away. But there is a restriction on what kinds of electron wave are possible. All must fit neatly inside the atom. Think of waves with one hump inside the atom. Or two. Or three. And so on. They *fit*. But waves with 1½ humps or 2.687 humps do not fit. This leads to a crucial distinction between an atom and the Solar System. Although in principle a planet can orbit at any distance from the Sun, an electron in an atom is most probably found only at *certain special distances* from the nucleus, corresponding only to *certain energies*.[18]

Immediately, this explains why atoms give out light of only certain energies, or wavelengths (the higher the energy, the shorter the wavelength). When an electron in an atom drops from a high-energy orbit to a low-energy orbit, it sheds its excess energy as light. The energy of this photon is equal to the difference in energy of the two states.

There is a twist. Isn't there always? An atom is a three-dimensional object. This means that an electron probability wave might be peaked not only at certain *distances* from the nucleus but also in certain *orientations*. Think of a globe. It takes two numbers to specify any location on the Earth. Similarly, it takes two numbers to specify an electron wave with a particular orientation in space. Add this to the number necessary to specify the distance of an electron from a nucleus and that makes a total of three quantum numbers.

The twist is therefore that an atom gives out light when it drops from *any* high-energy orbit to a low-energy orbit. And those orbits might differ not just in the distance of an electron from the nucleus but in the *orientation* of its orbit. Incidentally, the orientation of the electron wave – at least that of the outermost electrons – explains chemistry. An atom can join with another atom via its outermost electrons, which can be thought of as living on its exterior surface. And the locations on this surface where an atom can stick, or bond, with another atom are simply the locations at which the electron wave is biggest – that is, where electrons are most likely to be found.

The Pauli Exclusion Principle

There remains a puzzle, however. The electrons in an atom can occupy any of the permitted quantum waves, known as orbitals. But, in the real world, things have a strong tendency to minimise their energy. For instance, a ball, given a chance, will roll to the foot of a hill to minimise its gravitational energy. So, in an atom with more than one electron, why do they all not roll down the energy hill to the bottom? Why do they not crowd into the lowest-energy orbital, closest to the atom?

If this happened, atoms as we know them would not exist. For one thing, there could be no light. Photons are emitted only when an electron drops from one energy level to another, shedding its excess energy in the process. But, if all electrons were in the *same* state, at the *same* energy, there would never be any energy to be shed.

A more serious problem is that it is the *outermost* electrons that determine chemistry – how one type of atom joins up with other types of atoms to make molecules. For instance, some atoms have one outer electron, some two, three, and so on; and some atoms have outer electrons pointing in certain directions and other atoms in other directions, and so on. It is this that creates the huge variety of nature's atoms, from hydrogen, the lightest, to uranium, the heaviest. But, if all the electrons in every type of atom piled into the innermost orbital, all atoms would present pretty much the same exterior to the world. Instead of 92 naturally occurring kinds of atom, there would be only one. There would be no chemistry. No complexity. No us.

Once again, quantum theory rescues the atom – and the Universe – from such a stultifyingly dull fate. Recall that quantum ingredients, when combined in different permutations, spawn all sorts of novel and surprising behaviours. For instance, the mix of superpositions, electron spin and the law of conservation of angular momentum creates the madness of non-locality, of quantum particles influencing each other instantaneously when separated by impossible distances. Here is another mix: electron spin, the wave nature of electrons, and the fact that electrons are *indistinguishable*. The last is yet another new-under-the-Sun quantum property. Objects in the real world are always distinguishable – two similar cars by a scratch on the paintwork, for instance, or a slight variation in tyre pressure. But electrons are utterly indistinguishable. They cannot be marked. If two electrons are switched, there is no way to know that this has happened even in principle.

This mix of quantum ingredients spawns the Pauli Exclusion Principle.[19] In a nutshell, this edict says that no two electrons in an atom can share the same orbital. More specifically, no two electrons can share the same quantum numbers. A slight twist is that there is a fourth quantum number: spin. Remember, an electron in a magnetic field can either have a spin up or a spin down. So, the Pauli Principle says that no two electrons can share the same *four* quantum numbers.

The Pauli Principle stops electrons all piling on top of each other in the same orbital. It is why there are 92 types of naturally occurring atoms, not one. It is why there is variety in the world and you are here to read these words.

Because of the Pauli Principle, electrons arrange themselves in shells at successively greater distances from a nucleus. The first shell can contain a maximum of 2 electrons; the next 8; the next 18, and so on. Here are some examples . . . Take an atom with 6 electrons – it has 2 electrons in its inner shell and 4 in its outer shell. One with 12 electrons has 2 in its inner shell, 8 in the next shell and 2 in its outer shell. Immediately, it is clear why some kinds of atom behave similarly. For instance, lithium, sodium and potassium all have 1 electron in their outermost shell and so appear much the same to the outside world.

So there you have the origin of the world's stability and diversity. The wave-like nature of atoms prevents their electrons spiralling down to the nucleus in the merest split second. And the Pauli Exclusion Principle prevents the electrons piling on top of each other so, instead of just one kind of atom, there is a huge number of types. 'It is the fact that electrons cannot get on top of each other that makes tables and everything else solid,' said Feynman.

The Pauli Exclusion Principle applies to all subatomic particles with so-called half-integer spin – that is, $\frac{1}{2}$, $\frac{3}{2}$, $\frac{5}{2}$ units, and so on (quarks, by the way, have spin $\frac{1}{2}$ just like electrons). Such particles, known as fermions, are characterised by their enormous unsociability. Particles with integer spin – that is, 0, 1, 2 units, and so on – on the other hand, are gregarious. They do not obey the Pauli Exclusion Principle. This is why photons, which are bosons, are able to flock together in untold quadrillions to create the phenomenon of laser light.

It seems that there is nothing in the world that cannot be explained by quantum theory. It is the most successful physical theory ever devised. Inventions that exploit the ideas of quantum theory are estimated to account for 30 per cent of the GDP of the United States. Each and every one of us is a product of quantum theory. We live in a quantum world. But, although the quantum world is a magical world, there is little doubt that it is a mind-stretching world. 'If anybody says he can think about quantum physics without getting giddy', said Niels Bohr, 'that only shows he has not understood the first thing about it.'

16:

THE DISCOVERY OF SLOWNESS

Special Relativity

The velocity of light in our theory plays the part,
physically, of an infinitely great speed.

ALBERT EINSTEIN

When does Zurich stop at this train?

ALBERT EINSTEIN

Infinity is a number bigger than any other. If a body could travel at infinite speed, you would never be able to catch it up. Not only that but, no matter how fast you moved, the body would always appear to you to be infinitely fast, since your speed would always be negligible by comparison.

In our Universe, the role of infinite speed, for some reason, is played by the speed of light – 300,000 kilometres per second. No material body can ever catch it up. And no matter how fast you move relative to a source of light, or a source of light moves relative to you, the light will always appear to be travelling at 300,000 kilometres per second.

The remarkable fact that the speed of light – christened c by physicists – is doggedly unchanging was revealed by American physicists Albert Michelson and Edward Morley. In 1888, they measured the speed of light when the Earth, in its orbit around the Sun, was flying in the same direction as their light beam; and, six months later, when the Earth was moving in the opposite direction. To their consternation, they found that the speed of their light was the same in both cases. In fact, even had the Earth been orbiting the Sun at the truly enormous speed of half the speed of light, Michelson and Morley would still have measured the speed of light as c – not $(c + \frac{1}{2}c) = 1\frac{1}{2}c$; or $(c - \frac{1}{2}c) = \frac{1}{2}c$. What does the peculiar constancy of the speed of light mean? That question was answered by Einstein in his miraculous year of 1905.

The speed of anything is simply the distance it travels in a given time – for example, a car may travel 50 kilometres in an hour. So, for everyone to measure the same speed for a beam of light – no matter how fast they are moving or how fast the source of light is moving – something weird must have to happen to each person's measurement of distance and time.

We think of one person's interval of space – say, a metre – as being the same as someone else's, and one person's interval of time – say, a minute – as the same as another's. But this cannot be true if everyone is to measure the same speed for light. If the speed of light is the rock on which the Universe is built, space and time must be like shifting sand.

In fact, as Einstein realised, space *shrinks* and time *slows down* from the point of view of a moving observer. Or, to be more precise, if someone is moving relative to you, you see them shrink in their direction of motion and slow down as if they are moving through treacle.[1] 'Moving rulers shrink,' goes the saying, 'and moving clocks slow.'

Einstein had a more tongue-in-cheek – and, by today's standards, less PC – way of saying it: 'When a man sits with a pretty girl for an hour, it seems like a minute. But let him sit on a hot stove for a minute – it's longer than any hour. That's relativity!'

But what is it like from the point of view of the person flying past you? Well, they see *you* shrink in the direction of *your* motion; they see *you* slow down as if wading through treacle. This is because what you each see depends only on your *relative motion* – and both of you have the *same* relative motion.

This fact reveals the second foundation stone of Einstein's theory – in addition to the constancy of the speed of light – and explains why the theory is called relativity. Galileo, four centuries ago, was the first to realise that all people travelling at constant speed relative to each other see the same thing. Take, for instance, someone who throws a ball, which loops through the air to a friend who catches it. The ball will follow the same trajectory whether the thrower and catcher are on a beach or on a ship ploughing through the sea.

When Galileo maintained that all people travelling at constant speed relative to each other see the same thing, he specifically meant that they see the same laws of motion. Two and a half centuries later, Einstein simply extended Galileo's idea. It is not just the laws of motion that are the same, he claimed, it is *all* the laws of physics, including the laws of optics, which dictate that the speed of light is unvarying.

Think of the person moving past you at constant speed, and their space shrinking and their time slowing. From their point of view, you are moving with respect to them at the same relative speed – you are just moving backwards. So both of you see the same thing. That is the magic of relativity.

An obvious question is: why do we never see the weird effects of relativity – technically, time dilation and Lorentz contraction? Specifically, when someone runs past us on the street, why do we not see them shrink in the direction of their motion and slow down? The answer is that such effects are noticeable only for bodies flying past each other at speeds approaching that of light. But the speed of light is tremendously fast – about a million times faster than a passenger airliner. We do not see the effects of relativity because we live our lives in the cosmic slow lane. Relativity, in a sense, is the discovery of our slowness.

But, if we do not see the effects of relativity, how do we know that time really slows as we approach the speed of light? How do we know that space really contracts? The evidence is actually coursing through your body at this very instant.

Muons are subatomic particles, created about 12.5 kilometres up in the atmosphere when cosmic rays, high-energy atomic nuclei from supernovae, slam into atoms of the air. Like subatomic rain, muons shower down through the atmosphere. But there's the rub. A muon disintegrates after a characteristic interval of time.[2] The interval is very short – a mere 1.5 millionths of a second. By rights, therefore, none should travel more than about 500 metres down through the atmosphere before disintegrating. Certainly, none should reach the ground, 12.5 kilometres below.

But they do.

The reason is that muons are travelling at 99.92 per cent of the speed of light. From your point of view, they live their lives in slow motion. In fact, time passes 25 times slower for them than for you, which means they take 25 times as long as usual to realise it is time to disintegrate. When they do, they have already reached the Earth's surface.

But, of course, there is another point of view – that of the muon. From its angle, time is passing at its normal rate – after all, a muon is stationary *with respect to itself*, as are you. Instead, it sees *you* shrink in the direction of its motion – or, rather, *your* motion, since, from the point of view of a muon, it is the ground that is approaching at 99.92 per cent of the speed of light. But not only do you shrink, so too does the atmosphere. It shrinks to a mere 1/25th of its normal thickness. Which means the muons have time to get to the surface before they disintegrate.

Whatever way you look at it – from your point of view, where the muon's time slows down; or from the muon's point of view, where the atmosphere shrinks – the muon gets to the ground. It is one more example of the magic of relativity.

Space–time

But what would it be like if you, like a muon, could travel at close to the speed of light? For one thing, you would learn some profound truths about the world. You might think that relativity tells us that one person's interval of time is not the same as another's. It does. But, more specifically, it tells us that one person's interval of time is another person's interval of time *and* space. In other words, what one person sees as two separate events at the same location – say, two explosions – might appear to someone else as two events at different locations.

You might also think that relativity tells us that one person's interval of space is not the same as another's. But, actually, it tells us that one person's interval of space is another person's interval of space *and* time. In other words, what one person sees as two events happening simultaneously another person might see as two events happening at different times.[3]

But, if at speeds close to that of light, intervals of space morph into intervals of time and vice versa, then surely space and time cannot be fundamental things? Exactly. The fundamental entity, which becomes apparent only close to the speed of light, is *space–time*. It turns out that, in a low-speed world, we only ever see shadows of this seamless entity – a space shadow or a time shadow.

Here is an analogy. Imagine a walking stick suspended from its midpoint like a giant compass needle. It is in a square room with windows on two adjacent sides. Oh, and it is gloomy in the room so you cannot tell you are looking at a walking stick, just an object. You look through one window and you call what you see 'length'. Then you look through the adjacent window and you call what you see 'width'. Makes perfect sense. So far, so good.

Now imagine the walls of the room are on a turntable (this is not a simple analogy!). It turns. And you look through the windows again. To your surprise, you see that the length has changed. And so too has the width. It dawns on you that your labels of length and width were not sensible at all. The fundamental thing is the object – the suspended walking stick. But you have mistaken mere projections – shadows – of the object for the fundamental thing.

This is the way it is for space and time. The fundamental object is space–time. But we have mistaken mere projections of it – shadows – for the fundamental thing. It is not our fault. Our mistake becomes glaringly obvious only at speeds approaching that of light when space morphs into time, and time into space. Actually, in a deep sense, travelling close to the speed of light is like rotating our viewpoint – just as with the room on a turntable – so that we see different space and time projections of space–time.

It was not Einstein who had this insight but his former mathematics professor, Herman Minkowski, who famously called his pupil a 'lazy dog'. On later realising his mistake, Minkowski also recognised the key importance of space–time. 'From now on,' he said, 'space of itself and time of itself will sink into mere shadows and only a kind of union between them will survive.'

Mass is a form of energy

The speed of light is uncatchable by any material body. This is pretty amazing. After all, if we were really talking about infinite speed, it would be obvious that, no matter how hard and how long we pushed a body, it would never attain infinite speed. But the speed of light, though huge by human standards, seems *so* much smaller than infinity.

Well, since the speed of light is unattainable – the cosmic speed limit – something must happen as you push a body faster and faster. There must be some kind of resistance to your pushing, and the resistance must become infinite close to the speed of light so that no amount of pushing will ever get you there.

One property of a body provides resistance – its mass. In fact, that is how we define mass. A body that resists being pushed a lot, such as a loaded fridge, is said to have a big mass, while a body that resists very little, such as a feather, is said to have a small mass. See where this is going? If, as a body approaches the speed of light, its resistance grows, it must mean that it *gets more massive*.

But where is the extra mass coming from? There is a fundamental law of physics called the law of conservation of energy that says energy can be changed only from one form into another, and never created nor destroyed. For instance, electrical energy can be changed into heat energy in an electric fire; the chemical energy of your food can be changed into energy of motion in your muscles. But, if you are pushing and pushing the body and the energy you are putting in is not going into energy of motion, it must be going somewhere. Well, the only thing that is changing is the mass of the body. The energy you are putting in must be increasing its mass. But remember, energy can be transferred only from one type into another. *Mass must therefore be a form of energy.*

And, in fact, this is true. Not only did Einstein discover that space and time are mere facets of the same thing, he discovered that energy – energy of motion, sound energy, any energy you imagine – has an equivalent mass.

If you think all this is esoteric, with nothing much to do with you, think again. The quarks that contribute most of your mass are very insubstantial indeed.[4] In fact, they account for only about 1 per cent of your mass. This is explained by something called the Higgs mechanism. You may have heard of the Higgs particle, whose discovery was announced with a huge fanfare at the Large Hadron Collider near Geneva on 4 July 2012.[5] So where does the lion's share of your mass – the other 99 per cent – come from? The answer is relativity.

The quarks within the protons and neutrons of atoms are whirling around at close to the speed of light. This means they have enormous energy of motion. And this energy of motion, according to Einstein, has mass. It accounts for most of the mass of protons and neutrons – and, therefore, you. Without the effects of relativity, you would weigh less than 1 kilogram.

You may ask: why are the quarks whirling around at speeds approaching that of light? The answer is that they are in the grip of the enormously powerful strong nuclear force. A force field contains energy, which, according to Einstein, has mass. Ultimately, then, it is this gluon field that accounts for most of your mass. It does not matter how you look at it. Ultimately, something as mundane and everyday as your mass is inexplicable without the effects of relativity.

But Einstein showed that not only does energy have a mass but that mass has an energy associated with it. In fact, mass is the most concentrated form of energy known, and its energy is given by the most famous formula in all of physics, $E = mc^2$.

The formula applies both ways. Take subatomic particles circulating in opposite directions around the giant buried racetrack of the Large Hadron Collider. When the particles collide head on, their energy of motion can be converted into the mass energy of new particles, which appear out of the vacuum like rabbits out of a hat. But also – and this is the most shocking thing – mass energy can be converted into other forms of energy such as heat energy. This happens in a nuclear bomb, when a small amount of mass is converted into the tremendous amount of heat of a nuclear fireball.

You might think that relativity strips away our certainties about the world. But, in fact, it lifts the veil and reveals a deeper layer of reality beneath.

The world is complex, bewildering, ever changing. When we try to make sense of it, we are like shipwrecked mariners clinging to rocks in a turbulent sea. Physicists grab desperately for anything that seems solid and permanent. Specifically, things that are the same for everyone – that do not depend on a particular point of view.

Once upon a time, physicists believed space and time were the rocks of the Universe – that everyone would measure the same length of a given object; that everyone would measure the same interval of time between events. Einstein showed they were mistaken. What people measure depends on their point of view – specifically, how fast they are moving relative to each other.

Once upon a time, physicists believed mass was a rock of the Universe – that a body with a mass of 1 kilogram today would have a mass of 1 kilogram tomorrow and for all eternity. Einstein showed that they were mistaken. In an H-bomb, almost 1 per cent of the mass disappears, converted into other forms of energy, principally heat energy.

But what nature takes from us with one hand, it gives back with the other. Mass might not be the solid rock we thought it was – *but energy is*, with mass being merely one form of energy. Space and time might not be the rocks we thought they were – *but space–time* is.[6] Physics, it turns out, is the search for truths about the world that are independent of our point of view. Einstein, in lifting the veil of reality, showed us what is truly rock-like, truly invariant.

THE SOUND OF GRAVITY

General Relativity

If a bird-watching physicist falls off a cliff,
he doesn't worry about his binoculars; they fall
with him.

SIR HERMANN BONDI

One thing at least is certain, light has weight . . .
Light rays, when near the Sun, do not go straight.

ARTHUR EDDINGTON

Einstein's theory of relativity is a recipe for predicting what must happen to space and time in order for everyone to measure the same speed for a beam of light.[1] By 'everyone', Einstein meant people moving at constant speed relative to each other. A moment's thought, however, reveals that this is a very special circumstance. Very few bodies move with uniform speed. A car in traffic slows down and speeds up before coming to a halt at traffic lights. A rocket rising on a column of orange flame and white smoke gets ever faster until it attains the 29,000 kilometres an hour necessary to stay in orbit above the Earth.

So all Einstein had figured out in 1905 was what the world looks like from the point of view of atypical, or 'special', observers, moving at constant speed with respect to each other. For his next trick, he needed to figure out what the world looks like to typical, or 'general', observers, who are varying their speed with time, or accelerating, with respect to each other. Somehow, he had to turn his special theory of relativity into a general theory of relativity. It was a monumental task that would take him a decade of mental struggle, but it would cement his place in history as the greatest physicist since Isaac Newton.

In attempting to generalise special relativity, Einstein faced a serious problem. Not only does special relativity describe a special situation, it is *completely incompatible* with one of the great cornerstones of science – Newton's theory of gravity.

According to Newton, there is a force of attraction between any two bodies – for instance, the Sun and the Earth – that depends on their separation and on their masses. However, special relativity says that *all forms of energy* have an effective mass. If you heat a cup of coffee, for instance, its heat energy makes it marginally more massive than when it is cold. Consequently, all forms of energy must exert a gravitational force on each other, not just mass energy. It is *energy* not mass, as Newton believed, that is the source of gravity. Mass energy is simply the most familiar form.

And this is not the only incompatibility between special relativity and Newton's law of gravity. According to Einstein, light sets the ultimate cosmic speed limit. His theory therefore predicts that, if the Sun were to vanish suddenly – an unlikely scenario but imagine that it did – the Earth would not realise right away. For the time it takes gravity, travelling at the speed of light, to go between the Sun and the Earth, the planet would continue blithely in its orbit. Only after 8½ minutes would it realise that the Sun had disappeared and fly off on a tangent towards the stars.

Contrast this with the prediction of Newton's law of gravity. Two bodies feel an *instantaneous force* between them, which is synonymous with saying that the gravitational influence travels between them *infinitely fast*. Newton's theory therefore predicts that, if the Sun were to vanish suddenly, the Earth would notice immediately, in violation of Einstein's cosmic speed limit.

In concocting special relativity, Einstein had therefore inadvertently smashed one of the foundation stones of physics – Newton's law of gravity. He must have felt like a vandal who topples a beautiful building without the slightest idea how to build a replacement. But, just when he was despairing of ever finding a way, he had a brainwave. It concerned a simple observation that had been known about for centuries but whose significance no one before had recognised.

The seventeenth-century Italian scientist Galileo Galilei is supposed to have dropped a heavy mass and a light mass from the leaning Tower of Pisa and observed them hit the ground together. The same experiment, minus the complicating effect of air resistance, was later carried out on the Moon in 1972. Apollo 15 commander Dave Scott dropped a hammer and a feather together and, from the simultaneous puffs of moon dust, demonstrated that they hit the ground at the same instant.

Think for a moment how peculiar this is. If you were to take a small mass and a big mass and push both of them with exactly the same force, it is obvious that the small mass will gain the most speed. It is common experience that a big mass such as a fridge resists being moved more than a small mass such as a stool – it is the very basis of our *definition* of mass. But, when the force of gravity pulls a small mass and big mass groundwards, the two masses gain speed *at exactly the same rate*. In other words, the force of gravity goes up perfectly in step with the mass. Somehow, it *knows* to be bigger for a bigger mass. But how? It was Einstein's genius to think of a circumstance in which such an adjustment would come about perfectly naturally.

Imagine an astronaut in a rocket far from the gravity of any planet or moon. The rocket is accelerating, at 9.8 metres per second per second.[2] Since this is precisely the rate at which a falling body accelerates towards the Earth – 1g, in the jargon – the astronaut's feet are glued to the floor of his cabin just as if they were on the surface of the Earth. Now imagine that the astronaut holds a paperclip and a golf ball at the same height above the floor and lets them go simultaneously. Unsurprisingly, perhaps, they hit the floor at the same time, just as on Earth.

Now zoom out. Imagine you are floating outside the rocket with X-ray eyes that reveal to you the interior of the rocket (this not a realistic story). What do you see from your godlike point of view? The astronaut lets go of the paperclip and the golf ball and they hang motionless in space. How could they not do? The rocket, after all, is far from the gravity of any planet. But, as the two objects hang there, unmoving, the floor of the cabin *accelerates upwards to meet them*.

Now, recall that on Earth it was a complete mystery how gravity achieves the trick of adjusting its strength so that it pulls a big mass down at exactly the same rate as a small mass. But, in the rocket scenario, there is no mystery at all. Since it is the floor of the cabin that accelerates upwards to meet the motionless paperclip and golf ball, how could they not meet the floor at the same time?

But, wait a minute, the rocket scenario can explain gravity only if gravity *is the same as acceleration*. Exactly! Einstein's genius was to realise that the two things are completely indistinguishable. If the port holes of the rocket are blacked out and the vibration of the rocket is imperceptible, the astronaut experiences *exactly the same thing* as he would if he were in a blacked-out room on Earth. Gravity, Einstein realised, *is* acceleration.

Bizarrely, then, we are accelerating and we do not realise it. And, because we do not realise it, we have invented a force to explain what we experience: gravity.

It turns out that there is a point of view from which this fact is completely obvious, just as in the case of the rocket. But to appreciate it, it is first necessary to know a little background.

Gravity and time

The rocket thought experiment showed Einstein that gravity and acceleration are the same. If he could therefore find a theory of what the world looks like from the point of view of an accelerated person, he would automatically have a theory of gravity as well. *Two theories for the price of one.* But how? At this point, Einstein, still working as a Swiss patent clerk, had what he later called his greatest thought. 'The breakthrough came suddenly one day. I was sitting on a chair in my patent office in Bern. Suddenly, the thought struck me: If a man falls freely, he would not feel his own weight.'

Why was this a breakthrough? Well, since gravity and acceleration are the same thing, someone experiencing no weight – that is, no gravity – would not be accelerating. In other words, his situation would be described perfectly by special relativity, a theory that depicts what the world look likes to a non-accelerating observer. Not only had Einstein found that a theory of acceleration is one and the same thing as a theory of gravity, he had found the crucial bridge that connects it with special relativity, which he already had in his possession. A falling person feels no gravity and therefore his view of the world is described by special relativity.

A person accelerating with respect to him – that is, one experiencing gravity – could at each instant be assumed to be moving at constant speed. It was therefore possible to use special relativity to predict what his world looked like at one instant, then at the next instant, and so on.

Not surprisingly, since time appears to slow for someone moving with respect to you, time also appears to slow for someone accelerating with respect to you. But, since acceleration and

gravity are the same, this means time flows more slowly for someone experiencing stronger gravity.

In other words, *gravity slows time*.

Take two people working on the ground floor and top floor of
a building. The person on the ground floor is closer to the mass
of the Earth, and so experiences marginally stronger gravity.
Time therefore flows more slowly for them. If you want to
survive a long time, live in a bungalow.

This slowing of time, or time dilation, is fantastically tiny, and
you would need a super-precise atomic clock to show it. But, incredibly, in 2010, physicists at the National Institute of Standards
Technology in the US were able to show that, if you were to
stand one step lower than someone else on a staircase, time would
flow marginally more slowly for you.[3]

The slowing of time is appreciable, however, when gravity is
strong. And the strongest source of gravity we know of is a black
hole.[4] If you could hover near the edge, or horizon of a black
hole, time would flow so slowly for you that you would be able
to watch the entire future history of the Universe flash past your
eyes like a movie in fast-forward.

The hills and valleys of space

Back to that question – so far unanswered – of why, if gravity is
just acceleration, do we not realise we are accelerating?

Think of the rocket accelerating at 1g again. Imagine the
astronaut shines a laser beam across the cabin, from one wall to
the other, perfectly horizontally – at a height, say, of 1 metre
above the floor. What does he see? The beam strikes the far wall
of the cabin at a height of *less than 1 metre*.

This may seem peculiar. However, it is not unexpected. Although light is the fastest thing in the Universe, it nevertheless takes time to cross the cabin. And, during its flight, the floor of the cabin accelerates up towards it. The astronaut therefore sees the light beam curve downwards towards the floor. (For an acceleration of as little as 1g, the effect would be *very tiny* but it would be measurable by precision instruments.)

Two things. First, one of the defining characteristics of light is that it always takes the shortest path between any two points. The astronaut would therefore have to conclude from the trajectory of the laser beam that the shortest path in an accelerating rocket is not a straight line but a *curve*. Secondly – and this is the big thing – since acceleration is indistinguishable from gravity, the astronaut would have to conclude that the path of a light beam in the presence of gravity is a curve. In other words, *gravity bends light*.

Actually, Einstein had guessed, even before he came up with his general theory of relativity, that gravity bends the path of light. Special relativity, after all, predicts that all energy has an equivalent mass and therefore is affected by gravity (not to mention *exerts* gravity too). Since particles of light, or photons, possess energy, they have an effective mass and so should be bent by gravity. ('Photons have mass?!?' said Woody Allen. 'I didn't even know they were Catholic.')

Einstein's theory of gravity, however, adds a new and subtle twist to this light bending. The claim is that gravity and acceleration are equivalent. But, in the case of the rocket, this is not completely true. From the point of view of the astronaut, the two objects they release 'fall' towards the floor along parallel trajectories. However, this is not what happens if the same two objects are dropped on Earth. The reason is that gravity is always directed towards the *centre* of the Earth (in the extreme case of people living on opposite sides of the Earth, gravity pulls in *opposite directions*).[6] Because of this effect, the bending of a light beam by gravity is *twice as big* as naively expected.

Einstein's prediction of the gravitational bending of light was triumphantly confirmed on 29 May 1919 during a total eclipse of the Sun. Since in a total eclipse the glare of the Sun is blotted out by the Moon, it is the only time stars can be seen very close to the disc of the Sun.[7] As their light passes the enormous mass of the Sun on its way to the Earth, it should be deflected from its path by the gravity of the Sun. Sure enough, an expedition led by English astronomer Arthur Eddington to Principe, an island off the west coast of Africa, confirmed that the light bending was exactly as predicted by Einstein's general theory of relativity — twice the value expected from special relativity.

The bending of light by gravity provides the vital clue to answering the question: if gravity is acceleration, why do we not realise we are accelerating? Light, recall, always takes the shortest path between two points. Why, then, in the presence of gravity, does it follow a curved path?

Think of a hiker taking the shortest path through a range of hills. From the point of view of a high-flying bird, it is clear that the hiker does not follow a straight-line path. Instead, because of the undulations of the landscape, he pursues a tortuous *curved path*. The shortest path through a curved landscape is therefore not a straight line but a curve. See the parallel? If light follows a curved path in the presence of gravity, then it implies space in the presence of gravity is curved.

In fact, this is all gravity turns out to be: warped space, or, more precisely, *warped space–time*.

Nobody, before Einstein, suspected this. And no wonder. Space–time is a four-dimensional thing – it extends in the directions north–south, east–west, up–down and past–future. Since we are mere three-dimensional creatures, we are incapable of experiencing a four-dimensional reality directly.

Now, finally, we can understand why we are pinned to the surface of the Earth. There is no 'force' of gravity pinning us there – no invisible elastic holding us to the ground. Instead, space–time in the vicinity of the Earth is warped. We are at the bottom of a shallow valley of space–time. And we are accelerating downwards as surely as a ball heading to the bottom of a real valley. Only there is something in the way, stopping us: the ground. It is preventing us from falling. By pushing back, it is giving us the *sensation of gravity*.

It is no wonder that nobody before Einstein guessed that gravity was acceleration. Not only can we not see the valley of space–time we are in but the Earth's surface is obstructing our free fall.

Take another familiar example: the Moon, orbiting the Earth, is not held in the grip of the force of gravity as if attached to the Earth by a long piece of elastic. Instead, according to Einstein, the Earth warps the space–time around it, creating a valley. And the Moon flies around the rim of the valley like a roulette ball around a roulette wheel.

We realise none of this because we cannot directly experience the warpage of space–time. It took the genius of Einstein to guess its existence.

This analogy may help. You are a passenger in a car that makes a sharp turn. You feel yourself thrown outwards. And you attribute this to a force. If you know any physics, you will call it centrifugal force.

However, from the point of view of someone standing beside the road, no such force exists. You, the passenger, are simply continuing to move in a straight line. As the car rounds the bend, it is the *body of the car* that comes towards you. In the rocket example, the astronaut thinks he is experiencing gravity. But, from the point of view of someone outside (admittedly with X-ray eyes), there is no such force. He is just floating motionless. It is the floor of the rocket that comes up to meet him.

So, living on the surface of the Earth, we think there is a force of gravity because we do not realise we are accelerating and have hit something unmovable – the ground. We are accelerating because, unknown to us, space–time is curved.

There is no such thing as the 'force' of gravity. We are simply moving under our own inertia through curved space.

Einstein's theory of gravity – the general theory of relativity – can actually be encapsulated in a single sentence. It is due to the American physicist John Wheeler, who coined the term 'black hole'. 'Matter tells space–time how to curve,' said Wheeler, 'and curved space–time tells matter how to move.' That's all there is to it.

The devil, of course, is in the detail. General relativity is notorious for being easy to describe in words – and even in mathematical equations – while its implications in the real world are very hard to tease out.[8] Not only that but spotting the hand of general relativity in the outside world is extremely difficult. This is because its predictions tend to diverge from those of Newton's law of gravity only when gravity is strong. And gravity, on the Earth and in the Solar System, is very weak. If you do not think the Earth's gravity is weak, hold your arm out straight from your body. The Earth has a mass of 6,000 billion

billion tonnes yet the gravity of all that matter is incapable of pulling your arm downwards.

Gravity begets gravity

The general theory of relativity would have been found earlier had the evidence for it not been so subtle. But there was no need for it. Einstein's motivation was simply to generalise a theory he himself had concocted. This makes the general theory very unusual in the annals of science. Perhaps uniquely, it was not motivated by an observation of the world that did not fit the prevailing theory. Instead, it was one man's obsession.

Nevertheless, at the time Einstein was devising his general theory of relativity, there *was* an observation that contradicted a prediction of Newton's law of gravity. Few knew about it, let alone considered it important. It concerned the orbit of the planet Mercury.

Newton had discovered that the force of gravity between two masses weakens in a very particular way: if their separation is doubled, the force becomes four times weaker; if they are moved three times as far apart, it becomes nine times weaker; and so on.[9] Newton further showed, in a mathematical tour de force, that the path of a body, under the influence of such an inverse-square-law force, is an *ellipse*.[10] This explained the observation of the German astronomer Johann Kepler that the orbits of the planets around the Sun are not circular, as the Greeks maintained, but elliptical.

Actually, it is not quite true that each planet moves under the influence of an inverse-square-law force directed towards the Sun. In addition to being tugged by the Sun, each planet is tugged by *every other planet*, most significantly Jupiter, which is about 1/1000th the mass of the Sun. As a result, its orbit is not an unchanging ellipse but one that *very gradually* changes its orientation in space, or precesses, tracing out a rosette-like pattern. This is as true of Mercury as it is of any other planet. However, puzzlingly, Mercury's orbit precesses *above and beyond* that expected from the effect of the combined pull of all the other planets.

The mystery of Mercury's anomalous precession – the planet traces out a rosette that repeats roughly once every 3 million years – is explained by general relativity. According to Einstein, all forms of energy have an effective mass – heat energy, light energy, sound energy, and, crucially, gravitational energy. This means that, like all mass, it gravitates. In other words, *gravity creates more gravity*.

The effect of this is tiny and appreciable only where gravity is relatively strong – close to the Sun. Mercury is the planet closest to the Sun. Consequently, it experiences slightly stronger gravity than Newton would have predicted. Since a planet orbits in an exact ellipse only under the influence of a Newtonian inverse-square-law force, general relativity predicts that Mercury should show an anomalous precession, over and above that caused by the pull of the other planets. Einstein calculated the effect. Mercury, he discovered, should trace out a rosette in space that repeats about *once every 3 million years*. Exactly what is observed.

The fact that Einstein's theory can explain such an esoteric observation as the anomalous precession of Mercury was hardly likely to set the world on fire. What did that was the confirmation of the bending of light by gravity during the total eclipse of 1919. The observation – a confirmation by an Englishman of the prediction by a German, coming so soon after the catastrophe of the First World War – propelled Einstein into the scientific firmament. Instantly, he was hailed as the greatest physicist since Isaac Newton.

Light bending by the gravity of the Sun confirms that energy warps space–time while the anomalous precession of the orbit of Mercury confirms that all forms of energy – including gravitational energy – have gravity. But another prediction of Einstein's theory is that gravity slows down time. Long before it could be checked on Earth with super-sensitive atomic clocks, the effect was looked for in space – in the light emitted by white dwarfs.

The red shirt

A white dwarf is the endpoint of the evolution of a star such as the Sun. Having exhausted its heat-generating nuclear fuel, such a star continues to shine as its stored internal warmth gradually trickles away into space. A white dwarf packs the mass of the Sun into a volume no bigger than the Earth, making each sugar-cube-sized chunk of its matter roughly the weight of a family car. Crucially, such a dense object has a surface gravity about 10,000 times stronger than that of the Sun, which means that, according to Einstein, time should flow noticeably more slowly than on Earth.

For such an effect to be observable, the surface of a white dwarf must possess a clock that is easily visible. Remarkably, it does.

An atom of a particular element, such as sodium or iron, emits light of characteristic colours. These are unique to the element and are essentially its light fingerprint. Colour is merely a measure of how fast a light wave oscillates up and down. Such a regular oscillation is exactly like the ticking of a clock. And, sure enough, when astronomers observe a white dwarf and the light coming from a particular element on the star, they find that it oscillates *more sluggishly* than it would on Earth. In other words, the clock on the white dwarf ticks *more slowly*. And that slowing is precisely that predicted by Einstein.

Red light oscillates about half as fast as blue light. Consequently, the slowing down of the vibration of light shifts the light towards the red end of the spectrum.[11] This is why it is called the gravitational red shift.

And this red shift – or red *shirt*, as an article in the science magazine *New Scientist* once referred to it – is observed in other contexts too. For instance, when astronomers observe light from distant galaxies, they find that it too is oscillating more sluggishly than it would on Earth. This is the cosmological red shift. And it has the same cause as its counterpart on a white dwarf.

When we see distant galaxies, we see them as they were when the Universe was younger because their light has taken a long time to travel to us across space. When the Universe was younger, it was *smaller*. This is because the Universe is expanding, its constituent galaxies flying apart like bits of cosmic shrapnel in the aftermath of the big bang. Distant galaxies therefore inhabited a Universe where cosmic matter was squeezed to a higher density on average and had correspondingly stronger gravity than today. In such a Universe, time flowed more slowly than today, according to Einstein. We observe this in the sluggish oscillation of the light from distant galaxies – the cosmological red shift.

In physics, however, there is often more than one way to skin a cat. Another, entirely equivalent, way of viewing the red shift of light from distant galaxies and from white dwarfs is to say that, in climbing out of the strong gravity, the light *loses energy*. Since the energy of light is related to how fast it is oscillating, with high-energy light oscillating quickly, light that loses energy oscillates more sluggishly. It becomes *red-shifted*.

The sound of gravity

There remains one prediction of Einstein's theory of gravity that is yet to be confirmed directly: gravitational waves. In the general theory of relativity, space–time is not merely a passive canvas against which the drama of the Universe is played out. It is an active medium that can be warped by the presence of matter. In fact, it can be jiggled up and down too, creating a wave that propagates outwards like concentric ripples on a pond. This is a gravitational wave.

But space–time is not as elastic as the skin of a drum. It is a *billion billion billion* times stiffer than steel. This makes jiggling space–time to create strong gravitational waves very hard indeed. In practice, it requires some of the most extreme upheavals of matter in the Universe – the merger of two super-dense neutron stars[12] or black holes.[13]

But significant gravitational waves are generated by neutron stars and black holes *long before* they coalesce into one object – in fact, when they are still spiralling together. In 1974, this permitted an ingenious and elegant test of Einstein's theory. American astronomers Russell Hulse and Joseph Taylor discovered a system that consists of two neutron stars in orbit about each other. One is a pulsar, which, as it spins, sweeps a lighthouse beam of radio waves around the sky.

By carefully observing the binary pulsar, or PSR B1913+16, Hulse and Taylor determined that the two neutron stars are spiralling together, getting closer by about 3.5 metres each year. In the jargon, they are losing orbital energy. And, crucially, this lost energy is *exactly* the amount Einstein's theory predicts they should be radiating into space as gravitational waves. For this *indirect* proof of the existence of gravitational waves, Hulse and Taylor shared the 1993 Nobel Prize for Physics.

The race is now on to detect gravitational waves *directly*. In the beginning, people looked for them with huge suspended metal bars. The theory was that such a bar, when buffeted by a passing gravitational wave, would ring like a bell. But a myriad mundane terrestrial vibrations, such as waves sloshing on beaches thousands of kilometres away, can drown out such a minuscule signal.

In recent years, the technology of choice has been the laser interferometer which attempts to measure the deformation of space with rulers made out of laser light. Gravitational waves have the peculiar property that, as they pass, they simultaneously stretch matter in one direction while squeezing it in a perpendicular direction. Giant gravitational-wave detectors, each with two perpendicular arms to detect this effect, have been built in Europe and the US. The Laser Interferometer Gravitational Observatory (LIGO), for instance, consists of detectors in two different US states, and has perpendicular arms 4 kilometres long.

Physicists operating such detectors face apparently insurmountable difficulties. By the time gravitational ripples from even the most powerful astrophysical sources reach the Earth, they are enormously attenuated by distance. Experimenters are faced with detecting a deformation of space so tiny that it would change the distance between the Earth and the Sun by less than a tenth of an atomic diameter.

Likely sources of a strong enough pulse are not only a pair of neutron stars or black holes spiralling together but the birth of a black hole in the catastrophic collapse of the core of a star. The latter process is believed to occur when an extremely massive star detonates as a supernova.

Although there is an enormous amount of indirect evidence that black holes exist, there is no direct evidence because, by their very nature, they are very small and very black. However, in the formation of a black hole, the membrane, or event horizon, that surrounds it is expected to vibrate violently, generating copious gravitational waves. Crucially, just as the sound from a bell is unique to the bell, revealing its shape and size, the ripples in space–time spreading outwards from the birth of a black hole are expected to be an unmistakable signature of the event. The detection of the birth cry of a black hole will not only confirm Einstein's theory of gravity but at long last will provide the definitive proof of the existence of black holes.

The comparison of gravitational waves with sound waves is apt. For all of human history, we have obtained our knowledge of the Universe essentially from light – our sense of sight.[14] As far as the Universe is concerned, we have been stone deaf. Gravitational waves are the *sound of gravity*. Once we detect them, we shall at last be able to 'hear' the Universe.

18:

THE ROAR OF THINGS
EXTREMELY SMALL

Atoms

A physicist is just an atom's way of looking at itself.

NIELS BOHR

To see a World in a Grain of Sand
And a Heaven in a Wild Flower,
Hold Infinity in the palm of your hand
And Eternity in an hour.

WILLIAM BLAKE, 'Auguries of Innocence', 1803

Richard Feynman was arguably the most important American physicist of the post-war era. He won the Nobel Prize for devising the theory of quantum electrodynamics, which describes how light interacts with matter and, in doing so, explains pretty much every aspect of the everyday world. In *The Feynman Lectures on Physics*, Feynman asks, 'If in some cataclysm, all of scientific knowledge were to be destroyed, and only one sentence passed on to the next generations of creatures, what statement would contain the most information in the fewest words?' Feynman answers, 'All things are made of atoms.'[1]

The idea of the atom has an ancient history. Around 440 BC, the Greek philosopher Democritus picked up a stick or rock or it might have been a vase, and asked himself, 'If I could cut this object in half, then in half again, could I go on subdividing like this for ever?' To Democritus it was inconceivable that he could. Sooner or later, he reasoned, he would come to a grain of matter that could not be cut in half any more. Since the Greek for uncuttable was *a tomos*, Democritus called such an indivisible grain an 'atom'.

Democritus further postulated that atoms come in just a handful of different types. And by combining them in different ways it is possible to make a flower or a cloud or a newborn baby. 'By convention there is colour, by convention sweetness, by convention bitterness, but in reality there are atoms and the void,' said Democritus.

It was a remarkable leap of the imagination. The world around us looks bewilderingly complex. But this is an illusion, according to Democritus. Beneath the skin of reality things are simple.[2] Everything is made of a limited number of types of atom. Everything is in the combinations. Atoms, in short, are the alphabet of nature.

Democritus was led to his idea by the power of thought. But atoms, if they existed, were far too small to see with the naked eye. It took more than two millennia and the rise of science before indirect evidence was found for Democritus' idea. In a steam engine, for instance, steam pushes with a pressure on its container. If the container is fitted with a movable wall – a piston – this can then drive machinery such as a spinning machine or a train. The movement of the piston can be explained, scientists discovered, if steam consists of countless tiny atoms[3] flying about randomly through space. Their ceaseless drumming on a piston like raindrops on a tin roof creates a jittery force, which, smoothed out, we observe as pressure.[4]

'So many of the properties of matter, especially when in the gaseous form, can be deduced from the hypothesis that their minute parts are in rapid motion, the velocity increasing with the temperature,' said the nineteenth-century Scottish physicist James Clerk Maxwell. 'The relations between pressure, temperature and density in a perfect gas can be explained by supposing the particles move with uniform velocity in straight lines, striking against the sides of the containing vessel and thus producing pressure.'[5]

The behaviour of gases such as steam provides evidence of Democritus' idea that reality is composed of tiny grains of matter. But what about his idea that those grains also come in different types? Proof of this came from an unexpected direction.

For a long time, alchemists hoped it might be possible to turn base materials such as lead into precious stuff such as gold. Not only did they fail but they proved the opposite of what they set out to show. Some substances cannot, by any means, be broken down into simpler ones. In the late eighteenth century, the Frenchman Antoine Lavoisier guessed that such *elemental* materials are large collections of a single type of atom. Gold was an obvious element. But, over the years, chemists – the successors of alchemists – discovered many more. Today, we know of 92 naturally occurring elements, ranging from the lightest, hydrogen, all the way up to the heaviest, uranium – and we have even made heavier, artificial, elements such as plutonium.

By 1815, the English physician William Prout had noticed that most atoms appear to have a mass that is a whole-number multiple of the mass of a hydrogen atom. This led him to propose that atoms are actually made of smaller things – hydrogen atoms. Actually, as the atom was systematically broken apart in the late nineteenth and early twentieth century, it became clear that it was made of not one smaller thing but *three* smaller things. Protons and neutrons are close in mass to Prout's hydrogen building block while electrons are about 2,000 times lighter.

The picture that emerged gradually was of an atom as a miniature Solar System. At the centre, like a Sun, is a tiny nucleus, containing pretty much all the mass of the atom in the form of protons and neutrons (the exception being the lightest atom, hydrogen, whose nucleus contains only a proton). Around the nucleus, like planets around the Sun, there orbit electrons. The protons in a nucleus each have a positive electric charge and they are matched by an equal number of electrons with a negative charge. In fact, it is the force of attraction between opposite charges that keeps the electrons bound to the nucleus.

The planetary picture of the atom was deduced by the New Zealand physicist Ernest Rutherford in 1911. Rutherford's protégés, Hans Geiger and Ernest Marsden, had in 1909 fired sub-atomic bullets from the world's smallest machine-gun – a sample of radioactive radium – at a thin foil of gold. The picture of the atom at the time was of a Christmas pudding, with electrons studded like raisins in a sphere of positive charge. This predicted that Geiger's and Marsden's subatomic bullets – alpha particles – would fly through the gold atoms as surely as real bullets would fly though a cloud of gnats. To the astonishment of the two young experimenters, however, 1 in 8,000 *bounced back*. It took Rutherford two years to deduce that, contrary to the plum-pudding model, 99.9 per cent of the mass of the atom must be concentrated in a tiny nucleus, which 1 in 8,000 alpha particles had hit and bounced off.

One of the shocks of the planetary picture was of the incred-ible emptiness of an atom. A whopping 99.9999999999999 per cent is nothingness. If you could squeeze all the empty space out of all the atoms in all the people in the world, you could fit the human race in the volume of a sugar cube. The best image of the atom comes from the English playwright Tom Stoppard: 'Now make a fist, and if your fist is as big as the nucleus of an atom, then the atom is as big as St Paul's, and if it happens to be a hydrogen atom then it has a single electron flitting about like a moth in an empty cathedral, now by the dome, now by the altar.'[6]

The electrons, whirling far from the nucleus, represent the surface of the atom, where it makes contact with the world of other atoms. Their number – which is matched by the number of protons – therefore determines how an atom behaves; for instance, how it links with other atoms to make molecules. The lightest atom, hydrogen, has one proton in its nucleus and one electron circling; the second lightest, helium, two protons and two electrons; the third lightest, lithium, three protons and three electrons; and so on.

The neutrons in a nucleus carry no electric charge and play no role in determining how an atom presents itself to the world. Instead, they act as nuclear peacemakers, gluing the nucleus together via the strong nuclear force. Without their stabilising presence, the enormous electrical repulsion between protons would blast the nucleus apart.

It may seem that, with the atom turning out to be made of even smaller building blocks – protons, neutrons and electrons – Democritus' idea has gone out of the window. However, his proposal was merely that matter, ultimately, is made of *indivisible* grains. And it turns out that Democritus is right. The world *is* ultimately made of indivisible grains. It is just that they are not what we have chosen to call atoms. That is our mistake. The ultimate elemental building blocks turn out instead to be *subatomic* particles known as leptons and quarks.

Quarks and leptons

In fact, normal matter appears to be made of just four basic building blocks: two leptons and two quarks. The two leptons are the electron and the electron-neutrino. The electron is well known because most commonly it orbits in atoms, but the neutrino is less familiar, mainly because it is so amazingly unsociable. Although neutrinos are generated in prodigious quantities by the sunlight-generating nuclear reactions at the heart of the Sun, they interact with normal matter so rarely that they fly through the Earth as if it were transparent.[7] In fact, about 100 billion solar neutrinos are streaming through your thumbnail every second without you ever noticing. Eight and a half minutes ago, they were in the heart of the Sun.[8]

In addition to the two leptons, there are two quarks – the up-quark and the down-quark. These clump together in threes to make the proton and the neutron, with the proton consisting of two up-quarks and one down-quark, and the neutron two down-quarks and one up-quark. The existence of quarks was proved by essentially repeating Geiger's and Marsden's experiment of 1909 in which they fired alpha particles into atoms and saw that they were deflected by the atomic nucleus deep inside. Physicists instead fired electrons into protons. In experiments carried out in the late 1960s and early 1970s, the ricocheting electrons revealed the existence of three point-like particles deep inside: the quarks.

Although physicists are now certain that protons and neutrons are made of quarks, bizarrely it is impossible to knock one out and create a free quark. This is because of the peculiar behaviour of the strong nuclear force that glues together quarks. Not only is it super strong – Newton was right to say, 'The smallest particles may cohere by the strongest attractions' – it gets stronger the further apart are two quarks. It is as if they are joined by elastic that resists more the more it is stretched. Long before two quarks are free of each other, the energy put into stretching the 'elastic' is transformed into the mass energy of new particles, as permitted by the law of conservation of energy. Specifically, the laws of particle physics cause a quark–antiquark pair to be conjured into existence.[9] Experimenters must now separate two more quarks. But, in attempting to do that, they will create two more quarks, and so on.

But how do we know that the electron, neutrino, up-quark and down-quark are really nature's ultimate indivisible grains? The answer is because of the Pauli Exclusion Principle.[10] This quantum edict states that certain subatomic particles cannot share the same quantum numbers. In the case of the electron, this means that two electrons in an atom cannot share the same orbit (and spin). This ensures that electrons do not pile on top of each other, which would effectively make possible only one kind of atom rather than the 92 whose combinations create the variety of our world.

The Pauli Principle is a consequence of three things, one of which is that particles such as electrons are indistinguishable.[11] If two things are indistinguishable, it implies they have no substructure – otherwise the arrangement of their components could be used to tell them apart. The point is that leptons and quarks both obey the Pauli Principle. And the only way they can do this is if they are indistinguishable – that is, if they have no substructure and are truly nature's indivisible grains of matter.

So is that it? Ultimately, the world is made of just four building blocks – the electron, neutrino, up-quark and down-quark? Not quite. There is a twist. Isn't there always? For some mysterious reason, nature has decided to *triplicate* its building blocks! Instead of one quartet of particles, there are *three* quartets, each containing successively more massive versions of essentially the *same particles*. So, in addition to generation 1, which consists of the electron, electron-neutrino, up-quark and down-quark, there is generation 2, which consists of the *heavier* muon, muon-neutrino, strange-quark and charm-quark, and generation 3, which consists of the *even heavier* tau, tau-neutrino, bottom-quark and top-quark.

Bizarrely, neither of the two heavier families plays any role in the everyday world. In fact, since it takes a large amount of energy to create them, they were common only in the super-energetic fireball of the big bang in the first split second of the Universe's existence. When the muon – essentially a heavier version of the electron – was discovered in 1936, the American physicist 'I. I.' Rabi said, 'Who ordered *that*?' The same could be said of the all the duplicates of nature's four basic building blocks: 'Who ordered *them*?'

But how do we know there are not many more than three generations of fundamental building blocks? The answer comes from a surprising place: cosmology. Between 1 and 10 minutes after the birth of the Universe, the big-bang fireball was hot enough and dense enough for protons and neutrons to run into each other and stick together to make nuclei of the second heaviest element, helium. Remarkably, this primordial helium has survived until today and it can be observed throughout the Universe. Astronomers find that it accounts for about 10 per cent of all atoms. However, it turns out that, if there are many more generations of neutrinos, the gravity of their extra mass would have braked the expansion of the big-bang fireball, caus-ing the Universe to stay denser and hotter for longer so that it cooked up a different amount of helium. According to calcula-tions, a Universe with 10 per cent helium atoms is possible only if there are at most three or four generations of neutrinos. So there may be a fourth, even heavier, generation of fundamental building blocks still to be found. However, most physicists would bet against it.[12]

With three generations of particles, it would appear that there are a total of twelve fundamental building blocks – six quarks and six leptons. This is not quite all. There are also the forces that bind together the quarks and leptons – for instance, that glue the up-quarks and down-quarks into triplets to make protons and neutrons.

Force carriers

According to quantum theory, the forces arise from the exchange of *force-carrying particles*. Think of two tennis players hitting a tennis ball back and forth. As each player returns the ball, he feels the force of his opponent. Currently, physicists know of four fundamental forces: the electromagnetic force, which holds together the atoms in your body; the strong and the weak nuclear forces, which operate only inside the ultra-tiny domain of the atomic nucleus; and the gravitational force, which holds together planets, stars and galaxies. The electromagnetic force is carried by the photon; the weak nuclear by *three* vector bosons – the W^+, W^- and Z; the strong nuclear force by *eight* gluons; and the gravitational force by the graviton (though nobody has ever detected a graviton, and a quantum description of gravity in terms of such an exchange particle continues to elude physicists).

So now we are talking about twelve basic building blocks and thirteen force-carrying particles. Is that it? Well, actually, there is one other particle – the Higgs boson – which was discovered with much fanfare by the Large Hadron Collider near Geneva in Switzerland in 2012. This a localised lump in the Higgs field, a kind of invisible treacle that fills all of space and impedes the passage of the other particles, thereby endowing them with *mass*. Well, as ever, this is not quite the whole story.

Surprisingly, the quarks inside protons and neutrons are relatively light and account for a mere 1 per cent of the mass of normal matter, including you. This is what the Higgs explains. So where does the rest of their mass come from? The quarks are whirling about at close to the speed of light under the influence of the super-powerful strong nuclear force. It is their tremendous energy of motion that accounts for the missing 99 per cent of your mass since, as Einstein discovered, all energy has an effective mass.[13] Ultimately, that energy of motion – and therefore your mass – comes from the gluon fields responsible for the strong nuclear force.

So there you have it. There are twelve basic building blocks glued together by thirteen force-carrying particles with one extra particle connected to the field that gives all the other particles their masses. This quantum description of the fundamental building blocks and fundamental forces is known as the Standard Model and it is arguably the single greatest achievement of physics. Its major deficiency, however, is that it describes only three of the four fundamental forces of nature. Gravity, currently described by Einstein's general theory of relativity, remains stubbornly outside the fold.[14]

Twelve basic building blocks + thirteen force-carrying particles + the Higgs are rather a lot of fundamental particles. And, in fact, there are more – yet another twist. Each particle has associated with it an antiparticle, with opposite properties such as electrical charge or spin. A particle and its antiparticle are always born together, so the mystery is why we live in a matter-dominated Universe. The best guess of physicists is that, in the big bang, some lopsidedness in the laws of physics either favoured the creation of matter or preferentially destroyed antimatter. Incidentally, in addition to heavy particles (baryons) such as the proton and neutron, nature permits the existence of middleweight particles (mesons). Instead of being composed of a trio of quarks, these are made of just two quarks – a quark and an *anti-quark*.

So, to recap, there are twelve basic building blocks + thirteen forcecarrying particles + the Higgs + *all their antiparticles*. But physicists are always hoping to reduce the number. Ever since the late nineteenth century, when James Clerk Maxwell showed that the electric and the magnetic force are mere facets of a single *electromagnetic* force, physicists have been bitten by the unification bug. They are convinced that the four fundamental forces are merely facets of a single superforce, which reigned supreme in the high-energy conditions in the first moments of the big bang and which, as the temperature plummeted thereafter, repeatedly split into the forces we see today. In fact, in high-energy-particle collisions in the early 1980s, physicists actually witnessed the electromagnetic and weak nuclear forces merge back together into a single electroweak force.

Supersymmetry

In this spirit of unification, some physicists have suggested that the building-block particles, which are known as fermions, are merely different facets of the force-carrying particles, which are known as bosons.[15] A serious drawback of this elegant idea, known as supersymmetry, is that none of the known fermions seems to be the flipside of any of the known bosons! Undeterred, physicists have postulated that the supersymmetric partners of the known particles have very large masses and that current particle accelerators have insufficient oomph to create them in particle collisions.

If supersymmetry is right, it will show that fermions are the flipside of bosons. Unfortunately, it will do so only at the expense of generating a whole host of new particles! The hypothetical supersymmetric partner of the electron, for instance, is the selectron, and of the photon the photino. It might seem a high price to pay for unification. However, there could be a huge pay-off. The reason is that there is yet another twist to the story of the ultimate constituents of matter: dark matter.

Embarrassingly, the stuff made of atoms – the material you, me and the stars are made of and that science has focused exclusively on for 350 years – turns out to account for a mere *4.6 per cent* of the mass energy of the Universe.[16] A whopping 71.4 per cent is invisible dark energy – but that is not important here. The key thing is that 24 per cent of the mass energy of the Universe is in the form of dark matter, material that gives out no discernible light and whose existence is inferred only from the tug its gravity exerts on the visible stars and galaxies. The identity of the dark matter, which outweighs the Universe's visible stuff by a factor of more than five, is a mystery. However, one possibility is that it is made of hitherto undiscovered super-symmetric particles.

Supersymmetry is just a modern attempt to show that a range of phenomena is merely a presentation of different faces of a single, unified, phenomenon. The desire for such unifications, however, is on an inevitable collision course with Democritus' reductionist desire to show that reality is ultimately created by the permutations of a small number of basic building blocks. After all, if the reductionist programme ever succeeded in whittling down the fundamental building blocks to a single point-like fundamental particle, how could it have different faces? A point-like particle, by definition, looks the same from every viewpoint. There is, however, one way to avoid the conflict between unification and reductionism: if the fundamental building block is not a point-like particle. This is the proposal of string theory.

String theory

According to string theory, the fundamental building blocks of matter are one-dimensional strings of mass energy. These can oscillate like ultra-tiny violin strings, with ever more rapid, and therefore more energetic, vibrations manifesting themselves as heavier and heavier particles. One such vibration, for instance, would be the electron.

The strings are hypothesised to be fantastically small, typically a *million billion* times smaller than an atom. Probing such a tiny scale is way, way beyond our technological capabilities. It would require a particle accelerator to boost subatomic particles to an extraordinary energy since, according to quantum theory, the quantum wave associated with a particle is smaller the greater its momentum (or energy). Ultra-high energy is therefore synonymous with probing ultra-tiny scales – which is why the ultra-high-energy big bang echoed with the roar of things extremely small.

String theory has gained popularity because one particular string – a vibrating loop – has the properties of a graviton, the hypothetical carrier of the gravitational force. Thus string theory automatically incorporates gravity, which has proved the most difficult of the four forces of nature to unite with the others. The major drawback of the theory, however, is that, in order to reproduce the behaviour of all the fundamental forces, a total of ten dimensions is required – that is, six in addition to the four familiar ones. Proponents of string theory claim that the extra dimensions are not apparent because they are rolled up, or compactified, far smaller than an atom.

String theory is a possible candidate for a 'Theory of Everything'. Such a theory would explain all the fundamental building blocks and how they interact with each other via the fundamental forces in a single neat set of equations that could be scrawled on the back of a stamp – or at least on a postcard. It would, according to physicists such as Stephen Hawking, bring physics to a final and triumphant end. However, a Theory of Everything would not be what it is cracked up to be – and for two crucial reasons.

The first reason is that the Universe cannot simply be the inevitable consequence of a Theory of Everything. This is because a Theory of Everything, by its very nature, would be a *quantum theory*. In other words, such a theory would predict not what happens but merely the chances, or probabilities, of different things happening. Every time an electron is faced with the choice of going to the left of an obstacle or to the right of it, every time an atom is faced with the choice of spitting out a photon of light or not spitting one out, what it actually chooses is random. And this kind of thing has happened a myriad times since the big bang. The Universe we see around us today is not simply a consequence of a Theory of Everything but the consequence of a Theory of Everything *plus* a mind-blowingly large sequence of frozen accidents. Billions upon billions of other possible universes could have arisen – all from the same Theory of Everything. Ours is merely one – selected at random. 'Any entity in the world around us, such as an individual human being, owes its existence not only to the simple fundamental law of physics . . .' says American physicist Murray Gell-Mann, 'but also to the outcomes of an inconceivably long sequence of probabilistic events, each of which could have turned out differently.'[17]

The Theory of Everything, if we find it, will be a triumph of the human imagination. No doubt about that. But it will also reveal the limitations of the reductionist approach begun by Democritus two and a half millennia ago. We shall know the basic ingredients for a universe and the recipe. And that will be a fantastic achievement. But at the bottom of the recipe will be an instruction crucial to the success of the venture: cook for 13.77 billion years.[18]

The Theory of Everything has another limitation as well. The set of equations scrawled on the back of that stamp, or postcard, will describe how nature's fundamental building blocks of matter interact with each other via nature's fundamental forces. But it will not explain a newborn baby or a Shakespeare sonnet or why two people fall in love. How do these things arise? According to physicists, these phenomena emerge.

Emergence

A characteristic of the Universe – or at least our particular corner, the Earth – is that the fundamental building blocks combine together to make bigger building blocks, and these in turn link together to make even bigger building blocks, and so on. So, for instance, quarks and leptons combine with each other to make atoms. Atoms combine with each other to make molecules, including the mega-molecules of DNA. Molecules combine with each other to make gases and liquids and solids – and biological cells. Cells combine with each other to make plants and animals and human beings – and brains. And human brains combine with each other to make a global technological civilisation.

It is characteristic of this hierarchy, which spawns ever more novelty and complexity, that the laws that orchestrate how the building blocks at one level interact with each other give no hint of the laws that govern the behaviour of the building blocks at the next highest level, and so on. 'Life is not found in atoms or molecules or genes as such, but in organization,' according to American biologist Edwin Grant Conklin, 'not in symbiosis but in synthesis.'[19]

The whole is greater than the sum of its parts. For instance, a knowledge of how quarks glue themselves together to make the nuclei of atoms tells a chemist nothing about the behaviour of how atoms link together to make molecules. And a knowledge of how a single cell works tells a neuroscientist nothing about how 100 billion cells work in concert to make a human brain that can laugh, conceive a plan to send a man to the Moon or paint the *Mona Lisa*.

This is why, despite the fundamental status of physics, sitting at the bottom of the explanatory hierarchy of the world, it has not made redundant chemists or biologists or sociologists. At each level of complexity, new phenomena, described by new laws, emerge from the interaction of the building blocks at the level beneath. So, for instance, when large numbers of water molecules come together to make a water droplet, there emerges a property called wetness, which makes no sense for a single molecule of H_2O. And, when large numbers of atoms come together to make a pot of paint, there emerges a property called colour, which makes no sense for a single molecule of paint pigment.

Emergence might seem like magic but, really, it is not. Nothing is being added. It is more that information is being thrown away. Although it might be perfectly possible to predict the motion of a single molecule of gas flying through space, it is impossible to keep track of the countless quadrillions of gas atoms that make up the entire gas. So physicists ignore a vast amount of information. They approximate. They clear out some of those irrelevant details so that they can zoom out and see the big picture. This involves them inventing quantities such as pressure and temperature, which are averages of the behaviour of vast numbers of microscopic constituents. It is only when they zoom out intelligently in this way that they are able to spot correlations between the averaged-out parameters – new laws that govern their behaviour.

'The world of the quark has everything to do with a jaguar circling in the night,' wrote the Chinese-American poet Arthur Sze. But, in practice, connecting the two in a single explanatory framework is way beyond our twenty-first-century capabilities. We approximate because we do not have the mathematical ability to explain everything in terms of the behaviour of the most fundamental building blocks.

And we pile approximation upon approximation. And out of this process there emerge new laws. Ever more approximate laws. This is how we *understand* the Universe. This is how we make sense of a world that is far too complicated to be perceived in its entirety by the 3-pound lump of jelly and water that constitutes our puny ape brain. As Thomas Carlyle said, 'I don't pretend to

understand the Universe – it's a great deal bigger than I am . . .
People ought to be modester.'

What all this means is that a Theory of Everything, if it is
ever discovered, will not supplant all of science. Though it will
explain the interaction the fundamental building blocks of the
world, it will have nothing to say about flowers or sonnets or the
chuckle of a newborn baby.

19:

NO TIME LIKE THE PRESENT

Time

'What day is it?' asked Winnie the Pooh.
'It's today,' squeaked Piglet.
'My favourite day,' said Pooh.
A. A. MILNE, *Winnie the Pooh*

I can't talk to you in terms of time – your time
and my time are different.
GRAHAM GREENE, *The End of the Affair*

Imagine you look out of your window and see Normans, and, behind them, Romans, and, behind them, Egyptians. Crazy? No crazier than it is for astronomers looking out across the Universe with their telescopes. The further away a celestial object the *further back in time it is*.

Light travels at about 300,000 kilometres a second in a vacuum. But if, instead, it travelled at a mere *100 metres a century*, about a kilometre away you would indeed see William the Conqueror *still* invading England; about 2.2 kilometres away, Publius Scipio *still* battling Hannibal and his elephants; and not far from the horizon, about 4.5 kilometres away, the Pharaoh Khufu *still* making his weekly inspection of the building site of the Great Pyramid of Giza.

The reason all these events would still be visible is because, at 100 metres a century, the light bringing you news of them would *crawl snail-like* across the intervening distance. The point? Douglas Adams memorably observed, '*Space* is big. Really big. You just won't believe how vastly hugely mind-bogglingly big it is.'[1] What this means is that light, despite travelling *10 million billion* times faster than 100 metres a century, nevertheless *crawls snail-like* across the enormous expanses of the Universe.

Standing outside on a crystal-clear night, you see the Moon as it was 1¼ seconds in the past; the nearest star system, Alpha Centauri — and you need to live in the southern hemisphere to see this – as it was 4.3 years ago; and the Andromeda Galaxy – the most distant object visible to the naked eye – as it was when our *Homo erectus* ancestors were first venturing out onto the African savannah 2.5 million years ago.

With the aid of powerful telescopes, astronomers can drill back yet farther through cosmic time, revealing galaxies that lived and died long before the Sun and Earth were born. And, out at the very edge of the observable Universe, they can see the shimmering veil of the 'surface of last scattering',[2] 13.8 billion years back in time and the furthest it is possible to see with light.

What all this demonstrates is that time is not what we think it is. Because of the finite speed of light, time is inextricably bound up with distance. Or, as Einstein said, 'There is an inseparable connection between time and the signal velocity.' As we look outwards from the Earth, we might think we see the Universe as it is 'now'. But, actually, what we see are 'shells' of space at successively earlier times.

Telescopes drill through the onion-skin layers of cosmic time just as archaeologists dig through the dirt layers of terrestrial time. Astronomers, however, have the great advantage that they can actually *see* the past. Although they cannot know what the Universe looks like 'now', their compensation is that they can see the entire history of the Universe played out before their telescopic eyes.[3]

In our Universe, then, the concept of 'now' is meaningless. It is impossible know what it is like on Alpha Centauri at this moment since the light, carrying news of the star system, permits us to know only what it was like *a minimum of 4.3 years ago.*

The connection between time, space and the speed of light is as true on Earth as it is in the Universe. The crucial difference, however, is that terrestrial distances are far shorter. The light carrying an impression of the face of a friend you are talking with reaches your eyes in less than a billionth of a second. This is about 10 million times shorter than the briefest interval of time that can be perceived by your brain. Consequently, you notice no delay. The concept of 'now', which does not exist at all in the large-scale Universe, is on Earth in most circumstances a very good approximation. We can all safely assume that we are living in the same *present*.

Or can we?

Relativity and time

According to Einstein, the finite speed of light does more than simply *delay* news of events. Light is the cosmic speed limit and everyone, no matter what their speed relative to a source of light, measures exactly the same speed for a beam of light.[4] This can happen, as Einstein realised in 1905, only if the space of someone moving relative to you shrinks in the direction of their motion while their time slows down.

Einstein later generalised special relativity. According to his general theory of relativity of 1915, if someone is accelerating with respect to you – which is equivalent to *experiencing stronger gravity* – their time appears to slow down.[5] For instance, when astronomers look out across the Universe to a shell of space at an earlier epoch, the matter of the Universe at that time occupied a smaller volume than today – simply because space has been expanding since the big bang. With matter more concentrated, the Universe's overall gravity was stronger, and time flowed more slowly.

With time flowing at different rates for people moving relative to each other *or* who are experiencing different gravity, it is impossible for people to agree on what is past, present and future. In fact, the concept of a *common* past, present and future simply does not appear in Einstein's theory of relativity, our fundamental description of reality. The question then is: why do we have such a strong impression that it exists?

The answer is that the effects of relativity on time are appreciable only if two people are experiencing markedly different gravity or are moving relative to each other at an appreciable fraction of the speed of light. And, on Earth, all 7 billion of us are experiencing pretty much the *same* gravity and, even when flying in jet planes, moving relative to each other at less than a millionth the speed of light.

This is not true, by the way, for the Global Positioning Satellites, with respect to which electronic devices such as mobile phones calculate our location on the planet. In their elongated orbits, they swoop down towards the Earth before swinging back out into deep space. This means that not only do they speed up and slow down during each orbit but they also experience strong gravity close to the Earth and weaker gravity further away. As a consequence, the satellites do not experience a common past, present and future. And this must be taken into account by the program that computes our position relative to the GPS satellites. Relativity, it turns out, is not such an esoteric theory. It is an essential part of our everyday lives in the twenty-first century.

Still, we ourselves live our lives in the ultra-slow lane and in ultra-weak gravity where relativity would appear to have few consequences. Appearances, however, can be deceptive. Relativity, it turns out, still has a trick up its sleeve. And it has devastating consequences for our concept of time.

The map of space–time

Einstein showed not only that one person's interval of time is different from another person's interval of time. He showed that one person's interval of time is another person's interval of time *and space*. And that one person's interval of space is another person's interval of space *and time*. 'From now on, space of itself and time of itself will sink into mere shadows and only a kind of union between them will survive,' said Hermann Minkowski, Einstein's one-time mathematics professor.

Minkowski's union is space–time. 'The most important single lesson of relativity theory', says British physicist Roger Penrose, 'is that space and time are not concepts that can be considered independently of one another; they must be combined together to give a 4-dimensional picture of phenomena: the description in terms of *space–time*.'[6]

As lowly 3D creatures, we are incapable of experiencing 4D space–time in its full glory. All we can experience are *shadows* of 4D space–time, as Minkowski put it. And those shadows – space and time – change their magnitude depending on how fast we are moving relative to someone else. We might think we live in a universe with three dimensions of space and one of time but, actually, we live in a universe with four dimensions of space–time.

And herein lies the devastating problem for our concept of time.

Each of the four space–time dimensions has the *character of space*. Which means that space–time has the character of a *map* – a 4D map, granted, but a map none the less. And, just as New York,

Los Angeles and the Grand Canyon are locations on a terrestrial map, the big bang, the birth of the Earth and the end of the Universe are locations on the 4D map of space–time. Along with all the events of your life. What this means, according to Einstein, is that the past, present and future all exist *simultaneously*.

Disconcerting as this is to most people, it gave comfort to Einstein when, in 1955, his long-time friend, Michele Besso, died. In a letter to Besso's bereaved family (which they might not have entirely appreciated), Einstein wrote, 'Now he has departed from this strange world a little ahead of me. That means nothing. People like us, who believe in physics, know that the distinction between past, present, and future is only a stubbornly persistent illusion.'

But, if the past, present and future are only a stubborn illusion and in no sense do we actually *move through time*, why do we have such a strong sense that we do? In fact, why do we have such a strong sense that we are not only moving through time but moving through it in a *particular direction*? Why do we experience the past as L. P. Hartley's 'foreign country'?

For a long while – even before the advent of Einstein, who threw things into sharp focus – this was a complete mystery to physicists. The fundamental laws of physics do not prefer any direction of time. The law of gravity, for instance, could equally well allow the Earth to orbit the Sun in a backward direction. Despite this time reversibility, we emphatically cannot live our lives backwards, going from grave to cradle, growing younger with each passing year. Yet, incredibly, there is no explanation of why we feel we are moving through time – and in a particular direction – in fundamental physics. But there is such an explanation somewhere else – in thermodynamics.[7]

The arrow of time

If you were to show a picture of a castle and the same castle as a crumbled, vine-covered ruin, you would *know* that the derelict castle came later. Castles crumble. They do not uncrumble. The direction in which things decay, or become disordered, is the direction we associate with the direction of time. And it is the second law of thermodynamics that provides this 'arrow of time'.

There is a simple way of seeing this. Throw the fragments of a broken cup into the air. It is possible that the pieces come down to reassemble into an intact cup. However there is only *one* way this can happen, only one way a cup can be intact. Contrast this with the countless ways that the cup can come down in *even more broken pieces*. It is because there are overwhelmingly more ways that the cup can come down broken than intact – overwhelmingly more disordered states than ordered states – that cups break and do not unbreak. This is why time flows forwards but not backwards. It is why castles crumble but do not uncrumble, why coffee left in a cup grows cold rather than hot, and why people grow old rather than young.

And this is the way that the nineteenth-century Austrian physicist Ludwig Boltzmann formulated the second law of thermodynamics[8] – in terms of the number of possible ways in which the components of a body can be arranged and still be the body.[9] There is only one way for an intact cup. But trillions upon trillions for a broken cup. If all outcomes are equally likely, therefore, it is overwhemingly likely that a cup will stay broken, not leap back together as an intact cup. It is not utterly impossible – the second law of thermodynamics is different from fundamental laws of physics in not being cast iron but *statistical* – but the likelihood is you would have to wait many times the current age of the Universe to see such a bizarre thing happen.

So, even though our basic picture of reality – relativity – predicts that all of space–time is laid out like a map, and nothing actually moves through time, the thermodynamic arrow of time explains why we experience time flowing remorselessly in one direction only.

So, what is the ultimate origin of the arrow of time? Well, clearly, the Universe can get more disordered only if in the past it was *more ordered*. If it was already maximally disordered, it would have nowhere to go. So, the ultimate reason there is an arrow of time is that the Universe in big bang was in a highly ordered state.[10]

So maybe at last we are getting somewhere in understanding time. Although the concept of a common past, present and future appears nowhere in our fundamental description of reality – relativity – we nevertheless experience them because we live out our lives in the cosmic slow lane and in weak gravity. And, although relativity sets no direction for time, we grow old rather than young because the big bang was an unusual, highly ordered state.

But, actually, none of this really explains why we experience a present – why we focus our attention on the information most recently gathered by our senses. Why do we not have a *delayed present*, for instance, and focus on information that was collected, say, 10 seconds ago? Why do we not have *two presents*, and focus on data collected, say, 10 minutes apart? According to physicist Jim Hartle of the University of California at Santa Barbara, we are looking in entirely the wrong place when we look for an explanation in physics. Instead, we should be looking in biology.

No time like the present

Hartle thinks that, when life arose on Earth, organisms might have experienced time in a multitude of different ways. For instance, there might have been creatures with a delayed present, or two presents, or three presents, and so on. But imagine life for a tree frog with a delayed present, says Hartle. It sees a fly. It flicks out its tongue. But, because it is using out-of-date information, by the time its tongue is fully extended, the fly has long gone. Handicapped in this way, sadly, the frog eventually starves to death.

And this, says Hartle, is why we focus on most recently acquired information – why we have a 'now'.[11] Because it ensures our survival. Because all other ways of experiencing reality would have led to our extinction. Hartle therefore believes that, if there are other creatures in the Universe, they will experience time just like us.

Mysteries remain. For instance, our belief that time *flows* cannot possibly be true. After all, if something flows – such as a river – it changes with respect to something – like a river bank. If time really flows, then it must flow with respect to *something else*. There must be a second type of time, which is nonsense.

'Aside from Velcro,' says American humorist Dave Barry, 'time is the most mysterious substance in the Universe. You can't see it or touch it, yet a plumber can charge you upwards of seventy-five dollars per hour for it, without necessarily fixing anything.' Einstein's genius was not to get bogged down in what time *is* but to stick to what we can usefully say about it. 'Time', he said, 'is what a clock measures.' The great American physicist John Wheeler managed to encapsulate both Einstein's pragmatism and Barry's bafflement. 'Time', he said, 'is what stops everything happening at once.'

20

RULES OF THE GAME
The Laws of Physics

It is only slightly overstating the case to say that physics is the study of symmetry.
PHILIP ANDERSON[1]

Tyger! Tyger! burning bright
In the forests of the night,
What immortal hand or eye
Could frame thy fearful symmetry?
WILLIAM BLAKE, *Songs of Experience* (1794)

The world is not without order. The Sun comes up every morning. People grow old not young. An effect always follows a cause. There is a regularity to the loom of the world. 'Nature uses the longest threads to weave her patterns, so that each small piece of her fabric reveals the *organization* of the entire tapestry,' said American physicist Richard Feynman. That organisation hints that, behind the scenes, beneath the skin of reality, there are rules – laws – that orchestrate the world.

For much of human history, because of a belief that a Supreme Being was running things, people did not dig beneath the surface of reality. But Newton changed everything. Despite being a deeply religious man, he wanted to know the mind of God. He wanted to know the rules of the game by which the Supreme Being was orchestrating his creation. And, incredibly, Newton fathomed a universal law – one that applies at all places and at all times. The law of universal gravity quantifies how the force of attraction between two bodies depends on their masses and on the gulf of space between them. And it is remarkable in several ways.

First, formulating the law of gravity took a tremendous leap of the imagination, not to mention courage. At the time, the heavens were considered the domain of God. Even the Greeks believed that our planet was made of earth, air, fire and water – and that the heavens were made of an entirely different, and ethereal, 'fifth essence'. Newton dared to see heaven and earth as one and the same, subject to the same laws.

But how to tease out one of the laws? Newton's genius was to realise that a force of attraction exists between all masses, and that that force not only causes an apple to fall towards the Earth but the Moon to fall as well. The Moon, of course, does not look as if it is falling. However, the trajectory of a massive body in the absence of any force is a straight line, as spelled out in Newton's first law of motion (This not dissimilar from the law of cat inertia: 'A cat at rest will tend to remain at rest unless acted upon by some outside force such as the opening of cat food, or a nearby scurrying mouse.'[2]) Something must therefore be continually bending the Moon's path away from a straight line and towards the Earth. The something is the force of gravity between the two bodies. The Moon is falling towards the Earth as surely as an apple falls from a tree. The difference is that, as fast as the Moon falls, the Earth's surface beneath *curves away from it*. So it never gets any closer. Instead, it falls in a *perpetual circle*.

The Moon is falling far more slowly than an apple – it takes 27 days, after all, to circle the Earth. This enabled Newton, who knew how much further away the Moon was from the centre of the Earth than a falling apple, to deduce precisely how the force of gravity weakened with distance. It turns out it obeys an inverse-square law. In other words, the force of attraction between two masses becomes four times weaker if their separation is doubled; nine times weaker if it is tripled; and so on.

The details are not crucial. The key thing is that Newton had identified a universal law of physics. Today, we know that his law of gravity not only describes the fall of an apple from a tree and the motion of the Moon circling the Earth but also stars orbiting the centre of the Milky Way and galaxies orbiting within great galaxy clusters.[3]

One remarkable feature of Newton's universal law of gravity is that it is simple. Because the world around us looks bewilderingly complex, it might be expected that the laws that govern it are similarly complex – so complex that they are quite beyond the capabilities of our puny ape minds to grasp. But they are not. The Universe is so simple that, more than 350 years ago at the very dawn of science, one man was able to discern a universal law – one that operates at all times and in all places, from the beginning of time to the end of time, from one end of the Universe to the other.

But Newton not only discovered a simple and universal law, he discovered one that could be expressed *mathematically*. His compact formula summarises a huge range of phenomena.[4] Specifically, it relates the force of gravity between two bodies to their respective masses and their separation. Scientists ever since Newton have followed the great man's lead. And they have had ever more success in finding universal laws of nature, all of which are mathematical. This caused the Austrian physicist Eugene Wigner to remark on the 'unreasonable effectiveness of mathematics in the natural sciences'. Or, as the English physicist Paul Dirac put it, 'God is a mathematician of a very high order.'[5]

Most physicists still cannot quite believe that nature really dances to the tune of the symbols and mathematical relationships they scribble on a whiteboard or on a scrap paper. Time after time, they miss the messages in their own equations, just as Einstein missed the big bang in the equations he obtained that described the Universe. 'Our mistake is not that we take our theories too seriously,' said Nobel Prizewinner Steven Weinberg, 'but that we do not take them seriously enough.'

'The remarkable feature of physical laws is that they apply everywhere, whether or not you choose to believe in them,' says American astronomer Neil deGrasse Tyson.[6] 'Anyone who believes that the laws of physics are mere social conventions is invited to try transgressing those conventions from the windows of my apartment,' says American physicist Alan Sokal.[7] 'I live on the twenty-first floor.' But where do the laws of physics come from?

Symmetry

A clue – or, at least, an incisive way of thinking about this question – came from a German mathematician called Emmy Noether. In 1918, she made arguably one of the most significant discoveries in the history of science. She proved that the laws of physics are consequences of deep symmetries.

Symmetries are aspects of the world that are unchanged, or are invariant, under changes, or transformations.[8] 'A thing is symmetrical if there is something you can do to it so that, after you have finished doing it, it looks the same as before,' said the German physicist Herman Weyl. Take a starfish, which everyone knows has five arms.[9] A rotation of one-fifth of a turn leaves it looking exactly the same. Mathematicians say a starfish has five-fold rotational symmetry.

On the surface this may appear to have nothing to do with physics. However, nothing could be further from the truth. Noether's discovery is that symmetry gives rise to laws of physics.

Take the fact that, if you do an experiment today or next week, you will, all things being equal, get the same result. This time-translation symmetry spawns one of the most famous laws in physics – the law of conservation of energy, which says energy can neither be created or destroyed, merely morphed from one type into another – for instance from the chemical energy of petrol into the energy of motion of a car.

Then there is the fact that, if you do an experiment in London or New York, all things being equal, you get the same answer. This translational symmetry leads to the law of conservation of momentum, known instinctively by all snooker players. There exists a quantity known as the momentum, which is simply the mass of a body multiplied by its velocity. If a cue ball cannons into, say, a stationary red ball, the combined momentum of the cue ball and the red ball before the collision must be equal to the momentum of the cue ball and the red ball afterwards. This is the law of conservation of momentum.

Then there is a symmetry of two-dimensional space: rotational symmetry. If you do an experiment aligned in, say, a north–south direction, you get the same result as if the experiment is aligned east–west. This spawns the law of conservation of angular momentum. No need to know the details here. But it is the reason an ice skater spinning on the spot spins faster if she pulls in her arms.

But, of course, we live in a world with three dimensions of space and one of time. Actually, because of the constant speed of light, space and time are inextricably linked and we actually live in a universe with four dimensions of *space–time*. And rotational symmetries of the four space–time dimensions lead to the laws of Einstein's special theory of relativity. Still other symmetries of space–time spawn his general theory of relativity.

But this is by no means the end of the connection between symmetries and the laws of physics. Not by a long way. Quantum entities such as electrons can be imagined living in totally abstract spaces with dimensions that have a mathematical rather than a real existence. For instance, two quantum particles need six space dimensions to describe them; three particles nine, and so on. And, remarkably, symmetries of these abstract spaces spawn all the laws of quantum theory. In fact, so immensely powerful is Noether's theorem that all of physics can be seen as a consequence of deep, underlying symmetries. 'There is no law of physics that does not lend itself to most economical derivation from a symmetry principle,' said American physicist John Wheeler.

Noether's discovery that symmetry lies behind the basic laws of nature is the single most powerful idea in fundamental physics. Einstein agreed. On Noether's death in 1935, in a letter to the *New York Times*, he wrote, 'Fraulein Noether was the most significant creative mathematical genius thus far produced since the higher education of women began.'

But what are we to make of all these symmetries? Well, a striking aspect of all of them is that they are symmetries of *nothing*. That's right. The symmetries of normal space and time that spawn the conservation of energy and so on are also the symmetries of empty space and time – the symmetries of a void. And, it goes without saying, that the symmetries of abstract spaces that spawn the laws of quantum theory are also the symmetries of nothing. After all, they are symmetries of abstract mathematical spaces with no real existence. What all this means is that the laws of physics that orchestrate the Universe we live in are exactly the same as the laws of physics of an entirely empty universe. 'God made everything out of nothing, but the nothingness shows through,' said the French poet Paul Valéry.

All this is very suggestive. The ultimate cosmological question, after all, is: how did something come from nothing? The laws of physics are telling us that the Universe is closer to a state of nothing than we might have supposed. It is *structured nothing*. Maybe it was not as hard as we think to go from nothing to something?

As substances cool, they become less symmetric. Take water. It looks exactly the same at every location. However, when water freezes to make ice, the ice develops cracks and bubbles. It looks different at different locations.

Something like this is believed to have happened as the Universe expanded and cooled in the aftermath of the big bang. Originally, it was in a highly symmetric state. But, as the Universe cooled, more and more structure 'froze out', reducing the symmetry. Physicists talk of the symmetry being 'broken' so that today it is so well hidden that it takes the extraordinary ingenuity of physicists to uncloak it. This means that, when physicists recreate the conditions of the big-bang fireball by smashing together subatomic particles at the Large Hadron Collider, they recreate the more symmetric state of the primordial Universe, where it is easier to see the basic symmetries and deduce the fundamental laws.

Back to the water example again. At a lower temperature, the structured nothing of ice is simply more stable than the unstructured nothing of water. Could this be why we live in the Universe we do? The American physicist Victor Stenger thinks so. There is something rather than nothing, he says, because something is more stable than nothing. We, and everything around us, are simply patterns in the void.

PART FIVE: The cosmic connection

21

THE DAY WITHOUT A YESTERDAY

Cosmology

If you wish to make an apple pie from scratch you must first invent the universe.

CARL SAGAN

There is a theory which states that if ever anybody discovers exactly what the Universe is for and why it is here, it will instantly disappear and be replaced by something even more bizarre and inexplicable. There is another theory which states that this has already happened.

DOUGLAS ADAMS, *The Restaurant at the End of the Universe*

If you spent your whole life sitting on a chair in the middle of a field, you would find it hard – if not impossible – to create a mental picture of the Earth. Astronomers are similarly handicapped. They spend their whole lives pinned to the surface of a tiny ball of rock in an anonymous cosmic backwater. But, despite their overwhelming handicap, they have had remarkable success in concocting a picture of the cosmos. Not only do they know the content and extent of the Universe but they also have a pretty good idea of how it all came into being in the first place.

Nature has been kind to us. We do not live on a planet such as Venus shrouded in impenetrable clouds. We do not live in a star-choked region of the Galaxy such as the heart of the Milky Way where night is unknown. We have not appeared on the cosmic scene so late in the day that most of the stars have exhausted their fuel and sputtered out. Instead, using our Earthbound telescopes, we can see all the way to the Universe's distant horizon.

Previous generations would have killed for the kind of picture we have of our Universe. The Earth, along with a handful of other planets and moons and assorted leftover rubble from the formation of the Solar System, orbits the Sun. The Sun, in turn, orbits the centre of the Milky Way, a great pinwheel of about 100 billion stars turning ponderously in the night. At the Sun's location, about two-thirds of the way out towards the outer rim, it takes about 220 million years to complete a circuit, which means the last time the Earth was where it is at this very moment the dinosaurs were just beginning their 150-million-year reign.

The Milky Way, however, is but one island of stars, or galaxy, among 100 billion others. To get some idea of the scale of things, imagine the Universe is a sphere a kilometre across. It would be filled with 100 billion galaxies, each roughly the size of an aspirin, with the nearest galaxy, Andromeda, just over 10 centimetres away from us. Andromeda and a handful of the closest galaxies are bound to us by gravity. But all the other galaxies are fleeing from each other like pieces of cosmic shrapnel in the aftermath of an explosion.

The recession of the galaxies was discovered by American astronomer Edwin Hubble in 1929. An unavoidable consequence of this recession is that the Universe must have been smaller in the past. In fact, if the expansion of the Universe is imagined running backwards like a movie in reverse, a time is reached – about 13.8 billion years ago – when everything in creation was crowded into the tiniest of tiny volumes. This was the moment of the Universe's *birth*: the big bang.

One of the most profound discoveries in the history of science is undoubtedly that the Universe has not existed for ever, that it was born, that, in the words of the Belgian priest and mathematician George Lemaître, there was 'a day without a yesterday'.

Big bang

When the Universe was squeezed into a small volume, it must have been hot, for the same reason that air squeezed in a bicycle pump gets hot. The big bang was a *hot* big bang. The Universe was therefore born in a hot, dense state – a fireball. It has been expanding and cooling ever since and, out of the cooling debris, have congealed the galaxies we see about us, including the Milky Way.[1]

> Out of the cold and fleeing dust
> that is never and always,
> the silence and waste to come . . .
>
> This arm, this hand,
> my voice, your face, this love.
> JOHN HAINES[2]

The fireball of the big bang was like the fireball of a nuclear bomb. But whereas the heat of a nuclear fireball dissipates into the surrounding air in an hour, a day, a week, the heat of the big-bang fireball had nowhere to go. It was bottled up in the Universe, which, by definition, is all there is. Consequently, the heat of the big bang is still all around us today.

Although this cosmic background radiation was once blindingly bright, it has been greatly cooled by the expansion of the Universe since the big bang and no longer appears as visible light. Instead, it appears as microwaves, a type of light invisible to the naked eye but that can be picked up by your TV.[3] Tune your television between the stations. One per cent of the static, or snow, on the screen is the leftover heat from the big bang. Before it was intercepted by your TV aerial, it had been travelling for 13.8 billion years across space – and the very last thing it touched was the fireball of the big bang.

The cosmic background radiation is the most striking feature of our Universe. A remarkable 99.9 per cent of the particles of light, or photons, in the Universe are tied up in this afterglow of the big bang and a mere 0.1 per cent in the light of stars and galaxies. If we had eyes sensitive to microwaves rather than to visible light, we would see the whole of space glowing white like the inside of a giant light bulb.[4]

The afterglow of the big-bang fireball together with the expansion of the Universe are two powerful pieces of evidence that the Universe started out in a hot and dense state and has been expanding and cooling ever since.[5] One other major piece of evidence is that about 25 per cent of the mass of the Universe is in the form of helium, the second heaviest element. Starlight is a by-product of the fusion of the lightest element, hydrogen, into helium.[6] By estimating how much starlight there is in the Universe, astronomers can deduce that stars have converted only 1 or 2 per cent of the Universe's initial hydrogen into helium. So the helium had to be forged somewhere else.

The cores, or nuclei, of hydrogen atoms, being charged, repel each other ferociously. They can overcome their mutual aversion and stick together to make helium nuclei only if they slam into each other at high speed, which is synonymous with high temperature, and if they run into each other frequently, which is synonymous with high density. These twin conditions are believed to have been satisfied in the fireball of the big bang between about 1 and 10 minutes after the birth of the Universe. Calculations show that 25 per cent of the Universe's hydrogen should have been transformed into helium. And this is exactly the percentage of helium astronomers observe throughout the Universe.

The big-bang picture of the birth of the Universe provokes a host of questions. One of the most common is: where did the big bang happen? Here, the very term 'big bang', coined by English astronomer Sir Fred Hoyle on a BBC radio programme in 1949, sows seeds of confusion. After all, a bang, or explosion, happens at a particular location and the shrapnel flies outwards into pre-existing space. But the big bang did not happen at one location. It happened *everywhere*. And there was no pre-existing space. Space, along with matter, energy – and even time – were all created together in the big bang.

When we look out at the Universe and see all the galaxies fleeing from us, it does not mean that the big bang happened here on Earth. When astronomers say the Universe is expanding, all they mean is that *every* galaxy is receding from *every other* galaxy. In other words, if we could magically transport ourselves to another galaxy far across the Universe, we would also see all the galaxies fleeing from us. Everyone is at the centre and no one is at the centre because *there is no centre*.

An image often used to convey the idea is that of a cake with raisins baking in an oven. As the cake rises, every raisin recedes from every other raisin. No raisin is at the centre of the expansion. Of course, it is necessary to overlook the fact that a real cake has an edge – and imagine an infinite cake. But, then, all visual analogies of the big-bang expansion of the Universe provide at best a partial picture because it is *fundamentally unvisualisable*. The big bang, after all, happened in four dimensions of space–time – one dimension beyond what we, as lowly three-dimensional beings, can comprehend directly.

Many other questions provoked are by the big-bang picture of the birth of the Universe. What was the big bang? What drove the big bang? And, of course, what happened before the big bang? It is possible to answer all these questions. But only within

the context of a major – and it must be stressed, speculative –
extension of the basic big-bang model.

The extended big bang: inflation

The basic big-bang idea, for all its successes, contradicts our observations of the Universe in several serious ways. For one thing,
the cosmic background radiation comes to us more or less equally
from all directions in the sky. Or, to put it another way, everywhere in the sky has almost exactly the same temperature – 2.725°
above absolute zero.[7] This a problem because, if we imagine the
expansion of the Universe running backwards like a movie in
reverse to the time of the origin of the big-bang radiation, we
find that regions of the Universe that are today more than 1°
apart on the sky – twice the apparent width of the Moon – were
not in contact with each other.[8] Or, to be precise, there had been
insufficient time since the beginning of the Universe for any
influence – travelling even at the cosmic speed limit set by light –
to pass between them. Consequently, if one bit of the fireball
cooled down a bit faster than another, heat could not have
travelled to it from its surroundings to equalise the temperature.
The cosmic background radiation should therefore have an un-
even temperature across the sky. It should not, as is the case, have
the same temperature *everywhere in the sky*.

The bizarre explanation, which has been embraced by many
physicists, is that, in the Universe's first split second of existence,
it expanded far faster than at any time since – faster even than
the speed of light.[9] This period of inflation was so incredibly,
mind-bogglingly fast that the Universe doubled in size, and
doubled again, *more than 60 times over*. Inflation has been likened

to the explosion of an H-bomb compared to the puny stick of dynamite of the big-bang expansion that followed in its wake.

Inflation neatly explains why the temperature of the cosmic background radiation is the same everywhere we look in the sky. After all, if the Universe expanded far faster than we thought, it could have been smaller than we thought early on and yet still have reached its current size in 13.8 billion years. And, if it was smaller than we thought, then all bits could have been close enough to have exchanged heat, keeping the temperature of the Universe the same as it expanded.

Inflation, an idea from particle physics, was proposed in 1979 by the Russian physicist Alexei Starobinsky and independently in 1981 by the American physicist Alan Guth. Although the detailed physical mechanism underpinning the theory remains, frustratingly, obscure, inflation provides a majestic picture of the birth of our Universe and, most importantly, an explanation of what the big bang *was*.

This is the bizarre story now accepted by the majority of cosmologists. In the beginning was the inflationary, or false, vacuum. This was a weird, high-energy version of the true vacuum around us today.[10] For a start, it had repulsive gravity.[11] This caused the vacuum to expand, creating more vacuum, with more repulsive gravity, which caused the vacuum to expand even faster. Imagine you are holding a stack of banknotes between your hands and you pull your hands apart and more and more banknotes pop into existence. This is the way it was for the inflationary vacuum. Not surprisingly, physicists have dubbed inflation the 'ultimate free lunch'.

But the inflationary vacuum was intrinsically unstable.[12] Here and there, and totally at random, small patches disintegrated, or decayed, into normal, lower-energy vacuum. And, when this happened, the tremendous energy of the inflationary vacuum had to go somewhere. It went into creating matter and simultaneously heating it to a tremendously high temperature. *It made hot big bangs.*

Imagine a never-ending sea in which bubbles are appearing at random times and at random locations. Inside each bubble is a big-bang universe. One of those big-bang universes was *our Universe*.

Now it is possible to answer some of those nagging questions about the big bang. The big bang was not a one-off. It was merely a local event in an ever-expanding ocean of inflationary vacuum. It was driven by the energy of that decaying vacuum. And the big bang was not the beginning. Other big bangs have been going off like stuttering firecrackers across the length and breadth of the inflationary vacuum ever since the inflationary vacuum began, well, inflating.

As fast as bubble universes are created, they are driven apart. In fact, new vacuum is created far faster than it is eaten away, so inflation, once started, is unstoppable. It is *eternal*. But, even though inflation will continue into the infinite future, surprisingly this does not mean that inflation started in the infinite past. It must have had a beginning. The question of 'What happened before?' is therefore simply pushed back from the big bang to an earlier time. Quantum theory might come to the rescue, however, since quantum theory allows stuff literally to pop out of nothing. All that would have been necessary was for a tiny patch of false vacuum to pop into existence and begin inflating. Since a prerequisite of this happening is the existence of quantum theory, the question now becomes: 'Where did the laws of physics come from?'

The picture of the big bang as one among perhaps an infinite number of other bangs going off in an ever-expanding sea of vacuum is a remarkable one. But it is far from being bedded in firm theoretical ground. Inflation is an add-on, bolted onto the basic big-bang picture. It is not part of a single seamless theory of the Universe. And, worse, it is not the only thing bolted on.

The extended big bang: dark energy

The basic big-bang model not only predicts that the temperature of the cosmic background radiation should vary over the sky when it does not – something fixed by inflation – it predicts two other things that conflict with observations. For instance, it predicts that the expansion of the Universe should be slowing down. The galaxies, after all, are pulling on each other with their mutual gravity. It is as if they are connected by a vast web of elastic, dragging on them and hindering their headlong flight from each other. However, in 1998, physicists discovered that the expansion of the Universe, contrary to all expectations, is not slowing down. It is *speeding up*.

On the largest scales, another force must be operating in the Universe, overwhelming the force of gravity and driving apart the galaxies. This mysterious force appears to have switched on about 10 billion years ago and has been calling the cosmic shots ever since. The gaze of physicists has settled on the vacuum between the galaxies. They claim it is filled with dark energy. It is invisible. It fills all of space. And it has repulsive gravity. It is this repulsive gravity that is speeding up the expansion of the Universe.

Dark energy accounts for a 68.3 per cent of the mass energy of the Universe. Imagine how embarrassing it is to have overlooked the single biggest mass component of the Universe until 1998.

Dark energy could be an intrinsic energy of space predicted by Einstein's theory of gravity or it could have some other origin. Nobody knows. Physicists are pretty much at sea in explaining it. When quantum theory is used to predict the energy density of the cosmic vacuum – the dark energy – it comes up with an energy density of 1 followed by 120 zeros bigger than what is observed.[13] This is the biggest discrepancy between a prediction and an observation in the history of science. It does not take a genius to realise that some big idea is missing.

The dark energy, with its repulsive gravity, is reminiscent of the inflationary vacuum that speeded up the expansion of the Universe in its first split second of existence. The difference is that it was hugely more puny and nowhere near as short-lived. Nobody knows whether there is a connection between the dark energy and the inflationary vacuum.

Dark energy and inflation, however, are not the only two things that must be bolted on to the basic big-bang model to make it agree with what we observe. There is a third thing predicted by the big-bang model that is at odds with reality. In fact, it is quite a serious thing. The big bang predicts that *we should not exist*.

The extended big bang: dark matter

Recall that the galaxies, such as our own Milky Way, congealed out of the cooling debris of the big-bang fireball.[14] This was possible because the fireball was not completely uniform. The temperature of the cosmic background radiation is remarkably even all over the sky but it is *not totally even*. There are places where the temperature departs by a few parts in 100,000 from the average. The temperature undulations are believed to reflect the fact that some parts of the big-bang fireball were ever so slightly denser than their surroundings.

The small unevenness in the matter of the big-bang fireball is believed to have been caused by microscopic convulsions, or quantum fluctuations, of the inflationary vacuum in the first spilt second of the Universe. These were then magnified by the tremendous expansion of inflation. By about 379,000 years after the birth of the Universe, they had created slight bumps in the distribution of matter in the big-bang fireball. With slightly stronger gravity, these gathered matter about them faster, which boosted their gravity yet more. In a process akin to the rich getting ever richer, they grew remorselessly, eventually becoming the galaxies we see around us today.

It is a detailed and compelling picture. Only there is a problem. The 13.8 billion years since the big bang has not been enough time to assemble galaxies as big as the Milky Way. Not nearly enough. In short, we should not be here.

Undeterred, astronomers fix this problem by postulating that the Universe contains a vast quantity of invisible, or dark, matter, whose extra gravity speeded up the process of galaxy formation so it was completed within 13.8 billion years.[15] In fact, the dark matter in the Universe amounts to about 26.8 per cent of the mass of the Universe.[16] It outweighs the visible stars by a factor of more than five.

Nobody knows the identity of the dark matter. It could be in the form of fridge-sized black holes formed in the first split second of the Universe's existence.[17] Or it could be in the form of hitherto undetected subatomic particles. Certainly, theories of particle physics are not short of possible candidates. But the bottom line is that *nobody knows*.

To summarise, then, the basic big-bang picture must be supplemented by three bolt-ons: inflation, dark energy and dark matter. A figure of 68.3 per cent of the mass of the Universe is mysterious dark energy. Another 26.8 per cent is mysterious dark matter. That leaves a mere 4.9 per cent of the Universe made of ordinary matter – the stuff that you and I and the stars and galaxies are made of.[18] And, actually, we have only ever seen about half of that with our telescopes. The rest is ultra-hot hot gas floating around the galaxies that gives out little visible light.

To say that this is an embarrassing situation is an understatement. We have based the great edifice of our cosmological model on a mere 4.9 per cent of the Universe we have seen directly, whereas 95.1 per cent is made of invisible stuff whose identity eludes us. Imagine if Charles Darwin had tried to concoct a theory of biology knowing only of frogs but nothing of fish or birds or elephants.

Actually, it is not quite as bad as this. Dark matter and dark energy, to steal a phrase from Donald Rumsfeld, are 'known unknowns'. Astronomers, though hazy on the details, are confident that their overall picture is correct. Nevertheless, few would deny that there must be a deeper theory of the Universe out there, which unites inflation, dark matter and dark energy into a seamless whole.

Such a deeper theory might have to acknowledge one rather basic thing. Our Universe is not all there is. There might be other universes.

The multiverse

The key thing to remember is that the Universe was born 13.8 billion years ago. This means that we can see only those galaxies whose light has taken less than 13.8 billion years to reach us. Those whose light would take more than 13.8 billion years, well, their light is still on its way. Consequently, the Universe is bounded by a horizon – the light horizon. Think of it as the surface of a bubble. The bubble, centred on the Earth, contains about 100 billion galaxies, and is commonly known as the observable Universe.

But, just as we know there is more of the ocean over the horizon at sea, we know there is more of the Universe over the cosmic horizon. In fact, according to the theory of inflation, there is effectively an *infinite amount*. In other words, beyond the soap bubble of our observable Universe, are an infinite number of other soap bubbles. What is it like in them? Well, each had its own big bang – or, to look at it another way, a *portion* of our big bang. And, out of the cooling debris, congealed galaxies and stars – *different* galaxies and stars.[19] In other words, each bubble had a different history. 'Many and strange are the universes that drift like bubbles in the foam of the river of time,' said English science-fiction writer Arthur C. Clarke.[20]

There is a twist. Because the Universe is quantum, or grainy, there are only a finite number of possible histories for each bubble. Here is the reasoning . . .

According to quantum theory, the world at a microsocopic level is grainy, like a newspaper photograph. Ultimately, every-thing comes in indivisible chunks, or quanta. Energy comes in chunks. Matter does. Time does. And so does *space*. So, if we were able to look at space closely with some kind of super-microscope, we would see space resolve itself into indivisible grains. Think of it as a chessboard with squares of space. Now, if we run the expansion of the Universe all the way back to the beginning of inflation, we find that there were a mere 1,000 squares of space. The number is not relevant – although it is amazingly small. It is the fact that there is only a finite number of squares.

The seeds of galaxies turn out to be stuff on those squares. If a square contains energy that energy is the seed for a galaxy. But, just as there are only a finite number of ways to arrange the chess pieces on a chessboard, there is only a finite number of ways to fill the squares, some with energy, some empty of energy. In other words, the inflationary chessboard can create only a finite number of possible arrangements of galaxies, only a *finite* number of possible cosmic histories.

So we have a finite number of possible histories and an infinite number of locations for them to be played out. Consequently, every history occurs an infinite number of times. So there is an infinite number of copies of you whose lives until this moment have been exactly like yours. In fact, it is possible to calculate how far you would have to go to meet your nearest doppelgänger. The answer is roughly $10^{10^{28}}$ metres.

In scientific notation, the number 10^{28} is 1 followed by 28 zeros, which is 10 billion billion billion. Consequently, $10^{10^{28}}$ is 1 followed by 10 billion billion billion zeros. It is a tremendously big number. It corresponds to a distance enormously further than furthest limits probed by the world's biggest, most powerful telescopes. But do not get hung up on the size of this number. The point is not that your nearest double is at a mind-bogglingly great distance from the Earth. The point is that you have a double at all.

Don't believe this? Unfortunately, it is an unavoidable consequence of two things: a fundamental theory of the Universe and our fundamental theory of physics – quantum theory. If it is wrong, one or both of these must be wrong. This would not be an unusual state of affairs. 'Cosmologists are often wrong,' said the great Russian physicist Lev Landau, 'but they are never in doubt.'

22

MASTERS OF THE UNIVERSE

Black Holes

Now there's a look in your eyes,
Like black holes in the sky.
PINK FLOYD, 'Shine on you crazy diamond'

The black holes of nature are the most perfect
macroscopic objects there are in the Universe:
the only elements in their construction are our
concepts of space and time.
SUBRAHMANYAN CHANDRASEKHAR

Black holes are regions of space–time where gravity is so enormously strong that nothing, not even light, can escape. Probably, you think these celestial objects are esoteric and have no bearing on your everyday life. Nothing could be further from the truth. The birth of the Milky Way Galaxy, without which you would not be reading these words, might have been triggered by a black hole. Not only that but black holes reveal something about everyday reality that is so startling it is scarcely believable. Our Universe might be a giant hologram. *You* might be a hologram.

A black hole is a testament to the irresistible force of gravity, which, like the German football team, can be halted temporarily but always wins in the end. It triumphs because it is an attractive force between *every* piece of matter in the Universe and *every other* piece and nothing can nullify it. By contrast, the electromagnetic force – which holds together the atoms in your body – can be both attractive and repulsive and, on the large-scale at least, it is pretty much always cancelled out.

A black hole is a prediction of Einstein's theory of gravity, the general theory of relativity. It is cloaked by an event horizon, an imaginary membrane that marks the point of no return for in-falling matter and light. If an astronaut were able to hover just outside the event horizon, his time would slow down so much, as a consequence of Einstein's theory, that, in principle, it would be possible for him to look outwards and watch the entire future history of the Universe flash past his eyes like a movie in fast-forward.[1]

Inside the horizon, the distortion of time is so great that time and space actually swap places. This is why the singularity, the point at which in-falling matter is crushed out of existence at the centre of the hole, is unavoidable. It is exists not across space but across *time* so can no more be avoided than you can avoid tomorrow.

At the singularity, all physical quantities such as density sky-rocket to infinity. 'Black holes are where God divided by zero,' said the American comedian Stephen Wright. The singularity is an indication that Einstein's theory of gravity has been stretched beyond its limits and no longer has anything sensible to say. Almost certainly, a better theory – a quantum theory of gravity – will show that the singularity is not a singularity but instead just a super-high-density knot of mass energy.

'The black hole teaches us that space can be crumpled like a piece of paper into an infinitesimal dot, that time can be extinguished like a blown-out flame, and that the laws of physics that we regard as "sacred", as immutable, are anything but,'[2] said John Wheeler, the American physicist who popularised the term black hole.[3]

Although the general theory of relativity predicted black holes, Einstein never believed in their existence. This is not un-common in physics. Theorists often have difficulty overcoming their sheer disbelief that nature really dances to the tune of the arcane symbols they scrawl across a whiteboard. 'Our mistake is not that we take our theories too seriously,' as Nobel Prize-winning physicist Steven Weinberg observed, 'but that we do not take them seriously enough.'

Stellar black holes – which form from the catastrophic shrinkage of massive stars at the end of their lives – are, by their very nature, hard to spot. After all, they are very small and, well, very black. However, if a black hole is in a binary star system, we may see the telltale X-rays emitted by matter ripped from a companion star and heated to incandescence as it is sucked down into the hole. The first stellar-mass black hole, Cygnus X-1, was discovered by the Uhuru satellite in 1971. But, actually, something that would turn out to be far more significant in the black-hole story – and, crucially, more significant for *us* - was found eight years earlier.

Supermassive black holes

Quasars, discovered by Dutch-American astronomer Maarten Schmidt in 1963, are the super-bright cores of galaxies, blazing like beacons at the edge of the Universe. Because their light has taken most of the age of the Universe to reach us, they also blaze at the beginning of time. Typically, a quasar pumps out the energy of 100 normal galaxies such as the Milky Way but from a region smaller even than our Solar System. Nuclear energy – the power source of the stars – is woefully inadequate. The only process that can explain the prodigious energy output of quasars is matter heated to incandescence as it swirls down into a black hole. But not a mere stellar-mass black hole – a black hole with a mass of *billions of suns*.

For a long time after Schmidt's discovery, astronomers thought that such supermassive black holes were cosmic anomalies. They believed that such monsters powered only the badly behaved 1 per cent of galaxies of which the most extreme examples were quasars. But, over the past few decades, it has become clear that there is a supermassive black hole not only in the heart of these active galaxies but in the heart of pretty much *every* galaxy, including our own Milky Way. Most are quiescent, having gorged on and utterly exhausted their feedstuff of inter-stellar gas and ripped-apart stars.

The origin of supermassive black holes – unlike their stellar-mass cousins – is a mystery. Perhaps they are born when stellar-mass black holes collide and coalesce in the crowded heart of a galaxy. Or perhaps they form directly from the shrinkage of a giant pre-stellar cloud of gas. One thing is for sure: they grow extremely big extremely quickly. By the time the Universe was 500 million years old – a mere 5 per cent of its current age – there were supermassive black holes in existence that had already reached billions of solar masses.

But, although supermassive black holes are impressive on a human scale, they are minuscule on a cosmic scale. Not only are they tiny compared with their parent galaxies – even the biggest would fit easily within our Solar System – but they also have very small masses compared with the mass of the stars in their galaxies.

Despite this, they appear to control the stellar content and structure of their parent galaxies. For instance, the mass of the stars in the core of a galaxy is invariably about 1,000 times the mass of the central black hole, hinting that there is an intimate connection between a supermassive black hole and its parent galaxy. Consider for a moment how surprising this is. It is as if the growth of a mega-city such as Los Angeles were controlled by something as small as a single mosquito.

Jets

The means by which tiny supermassive black holes project their power over vast reaches of space are jets. Propelled by twisted magnetic fields in the gas swirling down to oblivion, these channels of super-fast matter stab outwards from the poles of the spinning black hole. They punch their way through the galaxy's stars and out into intergalactic space, where they puff up titanic balloons of hot gas – some of the largest structures in the known Universe.

In fact, such balloons of gas were the first hint that science got of the existence of supermassive black holes. In the 1950s, radio astronomers, using equipment adapted from war-time radar, discovered that the radio emission observed from some galaxies came not, as expected, from the central knot of stars but, mysteriously, from giant, radio-emitting lobe, on either side of the galaxy.

In the early 1980s, the thread-thin jets that are feeding the lobes were imaged for the first time by the 27 radio dishes of the Very Large Array in New Mexico. They mock our puny attempts at accelerating matter. Whereas the Large Hadron Collider near Geneva can whip a nanogram or so of matter to within a whisker of the speed of light, nature can boost many times the mass of the Sun each year to similar speeds along cosmic jets.

The jets control the structure of their parent galaxies because, in the inner regions where the jets are still fast and powerful, they drive out all the gaseous raw material of stars, snuffing out star formation. In the outer regions of galaxies, however, where the jets are slower, the jets have the opposite effect. As they slam into gas clouds, the concussion may trigger them to collapse under gravity to give birth to new clutches of stars.

But supermassive black holes, by starting and stopping star formation, do not merely sculpt galaxies. They might also determine the very *character* of the stars that form in them. Galaxies that contain the biggest supermassive black holes – so-called giant elliptical galaxies – appear to contain a much greater proportion of cool, red, long-lived stars, and there is evidence that the black hole might be responsible. Such red dwarfs spawn planets with few heavy elements such as carbon and magnesium and iron. Crucially, these are essential for life.

This has implications for our own Galaxy because, 27,000 light years away in the dark heart of our Milky Way, lurks a supermassive black hole 4.3 million times the mass of the Sun. Sagittarius A* might sound impressive but, actually, it is an insignificant tiddler compared with its 30-billion-solar-mass cousins in the cores of some quasars. Until recently, it was believed to be a mere coincidence that our Galaxy contains only a relatively small supermassive black hole. But is it? The giant elliptical galaxies that litter the cosmos might be chock-a-block full of planets but every last one of them might be a desert world, sterile and lifeless. The benign black hole in the heart of our Milky Way might be a large part of why we find ourselves here and not somewhere else. It might be a large part of why you are at this moment reading these words.

Did a supermassive black hole create the Milky Way?

Actually, we might be even more beholden to supermassive black holes than this. Most astronomers believe that galaxies give birth to supermassive black holes. But there is a contrary view and that is that supermassive black holes *give birth to galaxies*.

In this view, a giant gas cloud out in space shrinks catastrophically under its own gravity and, without forming any stars first, spawns a supermassive black hole. When its jets switch on, they stab outwards across space. If they happen to slam into an inert gas cloud floating in the void, the concussion causes the cloud to collapse, fragmenting into stars. In other words, it makes a galaxy.

This is no idle theoretical speculation. Astronomers know of a supermassive black hole floating in the void without a discernible galaxy of stars surrounding it. Extending from this naked quasar are two oppositely directed jets. And, at the end of one, is a newborn galaxy about the size of our Milky Way. It appears to have been triggered to form about 200 million years ago when the jet stabbed like a laser beam into a sleeping gas cloud. In the future, the supermassive black hole will fall into the heart of the galaxy it created and the galaxy birth process will be complete.[4]

If the idea that supermassive black holes zap galaxies into being is right, it is an extraordinary story how these objects have come in from the cold. Once they were thought to power only a tiny minority of anomalous active galaxies. Then it was discovered they exist in the heart of pretty much every galaxy. Now it appears they may actually *create* galaxies. You and I might owe our very existence to a supermassive black hole.

The holographic universe

But black holes, in addition to being essential to our existence, might also have something extraordinary to tell us about the Universe we live in – and indeed the nature of everyday reality. The Universe might be a hologram – a 3D representation of an

349

underlying 2D reality. *You*, without knowing it, might be a hologram.

Recall that a black hole is born when a massive star reaches the end of its life and shrinks catastrophically, crushing the star to a point-like singularity. The vanishing of a star in such a dramatic way was not a problem for physics until, in 1974, Stephen Hawking showed that, paradoxically, black holes are not completely black. They radiate into space so-called Hawking radiation.

Hawking imagined quantum processes going on just outside the event horizon. All the time, in the vacuum around us, sub-atomic particles and their antiparticles are popping into existence along with their antiparticles and then popping out of existence again. The energy to create such virtual particles is paid back quickly and so nature turns a blind eye. However, sometimes one particle of a pair falls into the black hole. The remaining partner, with no twin with which to annihilate, cannot pop back out of existence. No longer a fleetingly real particle, it now has a per-manent existence. The energy to create it has to come from some-where. And it comes from the gravitational energy of the black hole. Bit by bit, as the energy of countless particles of Hawking radiation has to be paid for, the hole loses its mass energy until, eventually, it vanishes, or evaporates.

The trouble with Hawking's discovery is that it implies that, when a black hole evaporates, all information about the star that initially shrunk to create the black hole – the type and location of all its atoms, for instance – will disappear too. This contradicts a fundamental edict of physics that information can never be created or destroyed.[5]

A clue to the resolution of the black-hole information paradox came from Israeli physicist Jacob Bekenstein. He discovered something profound about the event horizon: its surface area is related to the entropy of the black hole. In physics, a body's entropy is synonymous with its microscopic disorder.[6] But you do not need to know this. The crucial thing to know is that entropy is intimately related to information. A billion-digit number in which each digit is unrelated to the next has a high degree of disorder, or entropy; simultaneously, it contains a lot of information since the only way to convey it to someone is to tell them all billion digits.

This is the key clue to resolving the black-hole information paradox. In 1997, string theorist Juan Maldacena of Princeton's Institute for Advanced Study showed that it is in the horizon that the information that describes the star may be stored – as microscopic lumps and bumps. So, when the black hole radiates Hawking radiation from the vicinity of its horizon, the radiation has impressed on it information about the star, just as the radio waves from BBC Radio One have pop music impressed on them. So, when the black hole disappears, the song of the star is not lost at all. It is broadcast to the Universe as Hawking radiation. No information is ever lost.

But all this implies, incredibly, that a 2D surface – the horizon of a black hole – can store sufficient information to describe a 3D object – a star. This is exactly what the hologram on your credit card does.

This might seem an esoteric speculation about an esoteric type of celestial body. But, in the late 1990s, Leonard Susskind of Stanford University in California made a surprising and mind-blowing connection. The Universe, in common with a black hole, is surrounded by a horizon. It is a horizon in time rather than in space but it is a horizon none the less. So, reasoned Susskind, the information that describes the 3D Universe might be stored in the horizon of the Universe.

What this means is open to a wide range of interpretations. A conservative interpretation is simply that the Universe contains a lot less information than we imagined, meaning that the Universe is more like a crudely drawn sketch than a fine oil painting. A more extreme interpretation is that the Universe is truly a hologram – a 2D object stored on the cosmic horizon that creates the illusion of a 3D Universe. So, either we are living on that 2D surface, believing we are 3D, or our Universe is some kind of 3D projection of that 2D surface. You and I and everyone else might be living in a giant hologram. Black holes, far from being esoteric celestial objects, have the most profound implications for you and your everyday life. Black holes are indeed masters of the Universe.

ACKNOWLEDGEMENTS

My thanks to the following people who helped me directly, inspired me or simply encouraged me during the writing of this book: Karen, Neil Belton, Felicity Bryan, Manjit Kumar, Tim Harford, Ha-Joon Chang, Steve Russell, Reggie Kibbel, Lawrence Schulman, Nick Lane, Sue Bowler, Alex Holroyd, Chris Stringer, Steve Jones, Joanne Manaster, Adam Rutherford, Andy Coghlan, Carl Zimmer, Adrian Washbourne, John King, Chris Scarre, Brian May, Julian Loose, John Grindrod, Brian Chilver, Jose Tate, Karen Gunnell, Patrick O'Halloran, Jeremy Webb, Henry Volans, Simon Singh, Sarah Savitt, Tania Monteiro, Michele Topham, Valerie Jamieson, Roger Highfield, Alom Shaha, Peter Serafinowicz, Stuart Clark, Miles Poynton, Stephen Page, Silvia Novak, Jill Burrows.

PERMISSIONS

Extract from 'The Hollow Men' taken from *Collected Poems 1909–1962* © Estate of T. S. Eliot and reprinted by permission of Faber and Faber Ltd

Extract from 'Annus Mirabilis' taken from *High Windows* © Estate of Philip Larkin and reprinted by permission of Faber and Faber Ltd

Extract from *Hapgood* © Tom Stoppard and reprinted by permission of Faber and Faber Ltd

Every effort has been made to trace or contact all copyright holders. The publishers would be pleased to rectify at the earliest opportunity any omissions or errors brought to their notice.

NOTES

PART ONE: How We Work

I I AM A GALAXY: CELLS

1   Neurons are the longest cells in the human body. A single cell can stretch from your brain to the tip of your toe.

2   Lewis Thomas, *The Lives of a Cell*.

3   Typically, a bacterium can split into two bacteria in a couple of hours. At such a rate of doubling, after four days it can produce a million million offspring – enough to fill the volume of a sugar cube. After four more days, its descendants can fill a village pond. After another four days, the Pacific Ocean. In fact, in less than two weeks, a single bacterium can convert itself into a mass of bacteria equivalent to the mass of the Milky Way. Fortunately, this never happens. Just as the building of new houses requires a supply of bricks and mortar, the construction of new bacteria requires a supply of chemical building blocks. In practice, the supply is limited.

4   RNA is multi-talented. It can store information like DNA *and* behave like a protein – for instance, speeding up, or catalysing, chemical reactions. Since some RNAs can also replicate themselves, this has led to the idea that RNA pre-dated DNA. RNA's Achilles heel, however, is its fragility. Eventually, life found a more robust molecule for storing information, switching to DNA, which has a slightly different chemical backbone. In the 'DNA world', in contrast to 'RNA world', DNA recorded the recipes for making proteins, then sent out RNA copies of each recipe to the protein-making machinery of a cell. Thus proteins replaced RNA as catalysts and RNA was demoted to the role of a go-between.

5   In 1977, American biologist Carl Woese redrew the 'tree of life',

based on similarities between the DNA of organisms. At the base of Woese's tree are three trunks, or domains: *Bacteria*, *Archaea* and *Eucarya*. In the remote past, archaea bacteria split from bacteria. Only later did eukaryotes, which would spawn all multicellular creatures, including us, split from archaea bacteria. Archaea bacteria differ from bacteria in many ways, including the structure of their cell membranes. In fact, they have many things in common with eukaryotes, supporting the idea that they are the direct ancestor of the complex cells in our bodies.

6   The energy-generating chloroplasts inside the eukaryotic cells of a plant also look remarkably like free-living blue-green algae, or cyanobacteria. (Disc-like chloroplasts convert sunlight into chemical energy in a process called photosynthesis.) Cyanobacteria appear to have entered cells and set up home there about 2 billion years ago in an event that mirrors the swallowing of a bacterium by an archaea bacterium.

7   See Chapter 14, 'We are all steam engines: Thermodynamics'.

8   Lewis Thomas, *The Lives of a Cell*.

9   See 'Cell City', a BioPic production for the John Innes Centre and the Institute of Food Research, Norwich Research Park (http://www.biopic.co.uk/cellcity/index.htm).

10   Stephen Jay Gould, *Wonderful Life*.

11   Robert Brown also reported the curious dance of pollen grains in water. In 1905, Einstein realised this is due to the jittery bombardment of the grains by water 'atoms' (strictly speaking, molecules). Brown therefore has the distinction of helping to identify the fundamental building blocks of both physics *and* biology.

12   The DNA in a nucleus in a typical cell in your body, if unwound, would be about 2 metres long. Packing it into a nucleus about 6 thousandths of a millimetre across is like packing 40 kilometres of fine thread into a tennis ball.

13   Peter Gwynne, Sharon Begley and Mary Hager, 'The Secrets of the Human Cell'.

14   Adam Rutherford, *Creation*.

15   A fungus is a member of a large group of eukaryotic organisms that

includes microorganisms such as yeasts and moulds as well as mushrooms.

16   See Chapter 3, 'Walking backwards to the future: Evolution'.

17   Lewis Thomas, *The Lives of a Cell*.

2 THE ROCKET-FUELLED BABY: RESPIRATION

1   This is the fundamental recipe for a steam engine, the driving force behind all activity (see Chapter 14, 'We are all steam engines: Thermodynamics'). Energy goes from a high-temperature environment – and electrons in an atom with a lot of energy have a lot of energy of motion so can be considered *hot* – to a low-temperature environment – and electrons with little energy can be considered *cold*. In the process, the energy does *work*. In other words, it drives something against a force – in the case of a steam engine, a piston against air pressure.

2   The energy liberated by combining liquid hydrogen and liquid oxygen fuel is not quite enough to boost into space their combined weight *plus* that of the metal skin of a rocket. This is why a rocket is built in stages. A rocket, by dropping off a stage when it has climbed high into the air, makes itself lighter. Consequently, the fuel has an easier job of boosting it into space.

3   The electrons in an atom are arranged in shells, each with a maximum complement of electrons. Having a complete shell is hugely desirable. Hydrogen can achieve this by losing an electron (in fact, its sole electron); oxygen by gaining two electrons. This is why an oxygen atom grabs electrons from two hydrogen atoms. The state in which two hydrogen atoms lose an electron and an oxygen atom gains two electrons is the lowest-energy, desirable state, the equivalent of a ball lying at the foot of the hill.

4   Chemists talk of 'oxidation' and 'reduction' because, once upon a time, they did not know the precise details of what was going on in chemical reactions. In fact, an oxidising agent such as oxygen *grabs* electrons in order to reduce its energy, whereas a reducing agent such as hydrogen *donates* electrons to reduce its energy.

357

5   Proteins are large biomolecules used for a variety of purposes such as providing the scaffolding of cells and speeding up chemical reactions.

6   A proton, which is roughly 2,000 times more massive than an electron, is one of the two constituents of the core, or nucleus, of an atom. The other is a neutron. All atomic nuclei contain both particles, apart from the nucleus of a hydrogen atom, which contains only a proton.

7   Naively, it might be thought that an electron simply slams into a proton, driving it through a pore in the cell membrane. Actually, the electron changes the shape of a protein; it has one shape without the electron and another with the electron. Such shape changes force a proton across the membrane.

8   See Chapter 8, 'Thank goodness opposites attract: Electricity'.

9   Although the average person can survive without food for at most a month, there have been cases where people who are very obese have lived for a year on nothing but their own stored fat.

10   'The volume of blood passing through the human heart in an average lifetime would be enough to fill three supertankers,' according to @Qikipedia on Twitter.

11   Solar energy is not the only energy source of life on Earth. Some organisms exploit geochemical energy – for instance, the chemical reaction between molecular hydrogen (H_2) and carbon dioxide (CO_2). This is believed to have powered the very first living things on our planet.

3 WALKING BACKWARDS TO THE FUTURE: EVOLUTION

1   Useful traits are not only those that boost a creature's chance of surviving long enough to reproduce but also those that boost a creature's chance of getting the *opportunity* to reproduce if it survives that long. Such sexually selected traits include the peacock's tail – which makes a male attractive to a female – and a stag's antlers – which enables a male to out-compete other males for a mate.

2   Alfred Russel Wallace exempted humans from the process of natural

selection. He therefore avoided the controversy that surrounded Charles Darwin – and also the fame. Wallace's collected works – books, articles, manuscripts and illustrations – can be found at http://wallace-online.org.

3   The complete works of Charles Darwin can be found online at http://darwin-online.org.uk.

4   Actually, our Milky Way Galaxy turns out not to be at the centre of things but merely one among 100 billion or so others in our Universe. And there is a growing suspicion that our Universe itself is not special but merely one among countless others in a multiverse. So, seen in this context, Darwin is merely one of many scientists who have applied the Copernican principle, moving humans remorselessly from the centre of the world and revealing their insignificance in an indifferent, bewilderingly huge, and possibly infinite, cosmos. See Chapter 21, 'The day without a yesterday: Cosmology'.

5   See Chapter 1, 'I am a galaxy: Cells'.

6   The copying of DNA is made possible by a remarkable circumstance. *A* always pairs with *T*, and *G* with *C*. So, if a cell's double helix of DNA is split down the middle, it forms two complementary strands. *A*s floating about in solution automatically lock like jigsaw pieces to exposed *T*s; *T*s mesh with *A*s; *G*s with *C*s; and *C*s with *G*s. The result is *two* identical copies of the original DNA. No wonder that, when Francis Crick and James Watson discovered this in 1953, they rushed into the Eagle pub in Cambridge, England, and declared they had found the secret of life. See James Watson, *The Double Helix*.

7   In the 1960s, British biologist Lewis Wolpert ('Shaping Life', *New Scientist*, 1 September 2012) proposed that complex body plans of animals can be created by gradients in the concentration of chemicals across embryos. Depending on the local level of these compounds, known as morphogens, different genes get activated in different locations.

8   DNA stores information thousands upon thousands of times more compactly than the best current solid-state storage devices. In 2012, a team led by George Church of Harvard Medical School translated

into DNA a non-fiction book consisting of 53,000 words and 11 images. The team encoded the book in binary, using the bases *A* or *C* to represent a 'o' and *G* or *T* to represent a '1'. The book was the size of a typical bacterium's DNA. Cell division has already created *70 billion* copies of the book – 10 for every man, woman and child on Earth. All of them would fit in a single drop of water.

9   No one is absolutely sure why sex evolved since so many organisms do perfectly well without it. But one possibility is that it wrong-foots parasites, which are in a never-ending arms race with their hosts. A parasite can never become perfectly adapted to its host and kill a population if new variants of the host are continually thrown up. See Chapter 4, 'The big bang of sex: Sex'.

10 Michael Le Page, 'A Brief History of the Genome', *New Scientist*, 15 September 2012, p. 30.

11 Lewis Thomas, *The Lives of a Cell*.

12 Richard Dawkins, *The Blind Watchmaker*.

13 Gilbert Newton Lewis, *The Anatomy of Science*, pp. 158–9.

14 Author's telephone interview with Steve Jones.

15 Richard Dawkins, *The Greatest Show on Earth: The Evidence for Evolution*.

4 THE BIG BANG OF SEX: SEX

1   Winston Churchill's words, spoken in a radio broadcast in October 1939, actually referred to Russia. 'I cannot forecast to you the action of Russia. It is a riddle, wrapped in a mystery, inside an enigma . . .'

2   The first steps in the origin of life, according to a minority of scientists, occurred not on Earth but in interstellar space. Primitive bacteria were then ferried to the planet inside impacting comets. I wrote at length about panspermia – the idea that life on Earth was seeded from space – in the chapter entitled 'The Life Plague' of my book *The Universe Next Door*.

3   Lewis Thomas, *The Medusa and the Snail*.

4   Samuel Butler, *Life and Habit*.

5   Martin Luther, *The Table Talk of Martin Luther*, translated by William Hazlitt.

6   From time to time, there are scams to collect sperm from male 'geniuses' – scientists, artists, musicians and so on. Those behind them claim a woman using such a service – and it is best to skirt over the turkey-basting details – would give birth to genius children. However, this makes no biological sense. Even if the characteristics of a particular genius were determined by a certain gene sequence – and not by a gene sequence *plus* the influence of the environment – that gene sequence might not be inherited in its entirety by any offspring. Instead, they would share the genius's sequence *shuffled together* with the mother's sequence. Although it is certainly the case that a bacterial genius can beget another bacterial genius, this is unlikely to be the case for sexually reproducing organisms.

7   Leigh Van Valen, 'A New Evolutionary Law', *Evolutionary Theory*, vol. 1 (1973–76), p. 1.

8   Matt Ridley, *The Red Queen: Sex and the Evolution of Human Nature*.

9   Levi Morran, et al., 'Running with the Red Queen: Host-Parasite Coevolution Selects for Biparental Sex', *Science*, 8 July 2011, vol. 333, p. 216.

10   During DNA packaging, long pieces of the double-stranded molecule are tightly looped, coiled and folded so that they fit within the tiny nucleus of a cell. Eukaryotes achieve the necessary compaction by coiling their DNA around special proteins called histones to make a structure known as chromatin. They further compress the DNA through a twisting process called supercoiling. Most prokaryotes do not possess histones. Nevertheless, they use other proteins to bind together supercoiled forms of their DNA in much the same way as eukaryotes. Both eukaryotes and prokaryotes arrange this highly compacted DNA into chromosomes.

11   Ilea Leitch, et al., 'Evolution of DNA Amounts Across Land Plants (Embryophyta)', *Annals of Botany*, vol. 95 issue 1 (January 2005), p. 207.

12   See Chapter 1, 'I am a galaxy: Cells'.

13   Strictly speaking, you inherit slightly more DNA from your mother

than your father. This is because the energy-generating mito-chondria inside the egg have their own DNA, separate and distinct from the DNA of the whole cell. This is passed down exclusively from mother to child – and without any mingling with any other DNA. For this reason, mitochondrial DNA can be used to trace your ancestry.

14 See Chapter 6, 'The billion per cent advantage: Human evolution'.

15 The process of gene shuffling to make zygotes in meiosis crudely speaking takes one chromosome from each pair. This is done randomly and so each gamete can end up with any one of $2^{23} = \sim 10$ million possible chromosome combinations. This means that, for a man, the gametes will contain X, X, Y and Y. For a woman, they will all contain Xs. After sexual fusion of the gametes, the resulting zygotes will contain either XX, XX, XY or XY. Since Y determines maleness, 50 per cent will be male and 50 per cent female.

16 See 'The Origin of Sexual Reproduction', http://tinyurl.com/ca8sjwg.

17 See Chapter 5, 'Matter with curiosity: The brain'.

18 See Chapter 1, 'I am a galaxy: Cells'.

19 Most but not all sexually reproducing creatures have two sexes. Slime moulds, however, have thirteen. These single-celled amoeba-like creatures are neither animal nor plant but have things in common with both. Each sex can mate with all other sexes other than its own. (And you think you have problems finding and keeping a partner!) Incidentally, slime moulds, despite having no brain and being – well, slime – are nifty at finding their way out of mazes (see Ed Grabianowski, 'Why slime molds can solve mazes better than robots', www.io9.com, 12 October 2012, http://tinyurl.com/9ud95jx).

20 Philip Larkin, 'Annus Mirabilis', *High Windows*.

21 Richard Dawkins, 'The Ultraviolet Garden', Royal Institution Christmas Lecture No. 4, 1991.

5 THE MATTER WITH CURIOSITY: THE BRAIN

1 In Arthur Conan Doylr, 'The Adventure of the Mazarin Stone'.

2 In Kenneth Grahame, *The Wind in the Willows*.

3 Ambrose Bierce, *The Devil's Dictionary*.

4 Oscar Wilde, *De Profundis*.

5 Remarkably, if a sponge is minced up and its cells put in water, the cells will reconstitute themselves as a sponge once more.

6 If a charged atom or molecule is common in one location, such an ion will tend to move, or diffuse, to an area of lower concentration.

7 'The Origin of the Brain', http://tinyurl.com/d7sbhpk.

8 To be precise, the chemical messengers are contained in structures at the end of an axon known as terminal buttons. It is these that release them into the synaptic gap.

9 Interview with PBS, USA.

10 See Chapter 9, 'Programmable matter: Computers'.

11 Peter Norvig, 'Brainy Machines'.

12 Daniel Dennett, *Consciousness Explained*.

13 David Dalrymple, on leave from Harvard University, is aiming to build a complete simulation of the *C. elegans* nervous system. This will require first determining the function, behaviour and biophysics of each of the 302 neurons (Randal A. Koene, 'How to Copy a Brain', *New Scientist*, 27 October 20 12, p. 26). It is the first small step on the road towards a daring goal: the copying of a human brain into another material – for instance, the silicon of computers.

14 Edward O. Wilson, *Consilience*.

15 Sharon Begley, 'In Our Messy, Reptilian Brains'.

16 Spike Feresten, 'The Reverse Peephole', *Seinfeld* season 9 episode 12, 15 January 1998.

17 An outgrowth of a support cell known as a glial cell sheaths some neurons. The myelin sheath stops the electrical current of the axon leaking out into the surroundings just as plastic insulation stops electricity leaking out of the wires in your home. This is important if the current has to travel a long way – for instance, down the spine to the muscles of a limb. Myelin is white so neurons encased in it

are called white matter in contrast to the grey matter of the rest of the brain. People with multiple sclerosis, or MS, progressively lose the myelin sheaths around their white matter and so gradually lose the use of their limbs. Their thought processes, which are carried out in the grey matter, however, remain unaffected.

18 Gerald D. Fischbach, 'Mind and Brain', *Scientific American*, vol. 267 no. 3 (September 1992), p. 49.

19 Tim Berners-Lee, *Weaving the Web: The Past, Present and Future of the World Wide Web by its Inventor*.

20 Doris Lessing, *The Four-Gated City*.

21 George Johnson, *In the Palaces of Memory: How We Build the Worlds Inside Our Heads*.

22 Marvin Minsky, *The Society of Mind*.

23 James Watson, *Discovering the Brain*.

24 George E. Pugh (son of Emerson Pugh), *The Biological Origin of Human Values*.

6 THE BILLION PER CENT ADVANTAGE: HUMAN EVOLUTION

1 See Chapter 1, 'I am a galaxy: Cells'.

2 Hominin is a term that now includes chimpanzees.

3 Richard Dawkins, *The Ancestor's Tale: A Pilgrimage to the Dawn of Evolution*.

4 See Chapter 3, 'Walking backwards to the future: Evolution'.

5 See Chapter 13, 'Earth's aura: The atmosphere'.

6 Tutu has often used variations of this comment. Here is one: 'Tutu says apartheid, sin shattered when humans gather together' (ABP-news, 14 September 2006, http://tinyurl.com/nue9q6k).

7 See Chapter 12, 'No vestige of a beginning: Geology'.

8 As the Earth wobbles like a top, the tilt of its axis varies from 22.1° to 24.5° every 41,000 years. In addition, the elongation of the Earth's orbit varies every 100,000 and 400,000 years. These are collectively known as Milanković cycles. See Chapter 13, 'Earth's aura: The atmosphere'.

9 Richard Leakey and Roger Lewin, *Origins*.

10   Edward O. Wilson, *The Social Conquest of Earth*.

PART TWO: Putting Matter to Work

7 A LONG HISTORY OF GENETIC ENGINEERING: CIVILISATION

1   See Chapter 10, 'The invention of time travel: Money'.

2   While today's hunter-gatherer societies do appear to be egalitarian, they have been pushed by farmers into marginal habitats such as deserts. Ancient hunter-gatherers, by contrast, lived in environments with much more abundant animals and plants. It is possible, therefore, that they may not be directly comparable.

3   Sigmund Freud, *Civilisation and Its Discontents*.

4   Steven Pinker, *The Better Angels of Our Nature: A History of Violence and Humanity*.

5   Pat Shipman, 'Man's Best Friends: How Animals Made Us Human', *New Scientist*, 31 May 2011, p. 32.

6   According to a detailed analysis of a fossil dog skull carried out by a team led by Mietje Germonpré of the Royal Belgian Institute of Natural Sciences in Brussels (*Journal of Archaeological Science*, vol. 36 (2009), p. 473).

7   Jared Diamond, *Guns, Germs and Steel*.

8   Mark Twain, *Following the Equator*.

9   Heather Kelly, 'OMG, the text message turns 20', CNN, 3 December 2012, http://tinyurl.com/cgoakdg.

8 THANK GOODNESS OPPOSITES ATTRACT: ELECTRICITY

1   The force, like gravity, weakens according to an inverse-square law – that is, if two bodies are moved twice as far apart, the force is four times weaker; three times farther apart, nine times weaker; and so on.

2   To be precise, the electric force between an electron orbiting a proton in an atom of hydrogen, the lightest element, is about 10^{40} – that is, *10,000 billion billion billion billion* – times stronger than the

gravitational force between them. Crudely speaking, this means at least 10,000 billion billion billion billion atoms must clump together before gravity – which is cumulative – can overcome the electrical repulsion between those atoms. This corresponds to a body about 600 kilometres across if made of rock and about 400 kilometres if made of ice, which is less stiff than rock and more easily crushed by gravity. Above this threshold, gravity can overwhelm the electrical force, pulling every component of a body as close to the centre as possible and creating a sphere. This explains why all bodies in the Solar System smaller than the threshold are potato-shaped while all bodies bigger are spheres like the Earth.

3   Why, if the protons in an atomic nucleus and the electrons in orbit around the nucleus attract each other with such a tremendous force, do atoms not simply shrink to zero size? The answer is that the fundamental building blocks of matter have a peculiar wave-like nature, and waves are fundamentally *spread-out* things. It is because the electron wave needs a lot of elbow room that the electron violently resists being squeezed too close to a nucleus. Without this quantum effect, atoms would not exist (see Chapter 15, 'Magic without magic: Quantum theory').

4   As much energy would be needed to remove all the electrons from a mosquito as would be liberated by its explosion. This is because the law of conservation of energy dictates that energy cannot be created or destroyed, merely changed from one form into another.

5   Benjamin Franklin advocated a single fluid model of electricity in which objects with a deficiency of the fluid had a negative charge and objects with a surplus a positive charge. He was right about the single fluid but wrong on the rest. Most commonly, the fluid – electricity – is made of negatively charged electrons – and it is a *surplus* of them not a deficiency that makes a body negative. By the time this was realised, however, an electric current had already been defined as a *flow of positive charge* and the idea was so well established that no one wanted to change it. Because of a historical accident, therefore, an electric current flows in the opposite direction to the actual flow of electrons. Incidentally, Franklin coined the terms

'negative' and 'positive'. He was also responsible for other electrical terms such as 'battery' and 'conductor'.

6   Fast photography actually reveals not a single discharge in each lightning bolt but many, each lasting only a millisecond or so. The first, known as the leader, is actually from the ground upwards to the cloud. It is followed by alternating discharges from the cloud to the ground. To the human eye all the discharges blend into a single strike, though it may appear to flicker.

7   Typically, a single stroke of lightning contains enough energy to light 250 homes for an hour. Most of the 100 lightning bolts that occur each second across the world are in the tropics. This is because the key to creating charge imbalances is updrafts of air. The equatorial region has the most rising hot air since it intercepts the lion's share of solar heat.

8   Edison famously tested hundreds of possible filament materials to find one that glowed brightly without disintegrating. 'I have not failed,' he said, describing his method. 'I've just found 10,000 ways that won't work.'

9   The mechanism by which electric charge becomes separated in a thunderstorm is very similar to rubbing a balloon against a nylon sweater. Moist air, caught in a violent updraught, rises and cools to form ice crystals. In the turbulent air, the crystals run against each other, friction transferring electrons between them and building up a large charge imbalance between one cloud and another or between a cloud and the ground.

10   At first sight it appears impossible that a scrap of paper that is an equal mix of negative and positive charges can be attracted by a charged balloon. It happens in the following way. Say, for instance, the balloon is negatively charged. Its negative charge repels the negative electrons in the surface of the paper, causing them to move deeper into the paper. This leaves the surface of the paper with a net positive charge. It is this positive charge that is attracted by the negative charge of the balloon. A charged body, like the balloon, is said to separate, or polarise, the charge of an uncharged body, like a scrap of paper.

11   In the modern, or quantum, picture, force fields are the result of the exchange of force-carrying particles. The electric force is a consequence of the exchange between charged particles of virtual photons, a type closely related to the photons of light.

12   'I encountered a wonder . . . as a child of four or five years when my father showed me a compass,' said Einstein. 'That this needle behaved in such a determined way did not fit into the way of incidents at all . . . There must have been something behind things that was deeply hidden.'

13   Just as electric charge comes in two types – positive and negative – magnetic charge comes in two types – north poles and south poles. And, just as like charges repel and unlike charges attract, like poles repel and unlike poles attract. But here the similarity ends. Although it is perfectly possible to have an isolated negative charge or an isolated positive charge, nobody has ever observed either an isolated north magnetic pole or an isolated south magnetic pole. A north pole *always* comes with a south pole.

14   From Walter Elsasser, *Memoirs of a Physicist in the Atomic Age*.

15   The wire must be a conductor. This is a material such as copper or silver, whose atoms possess loosely bound electrons that can be detached and driven through the material by an electric field.

16   Nuclear power, as Terry Jones of *Monty Python* remarked, is 'a very silly way to boil water' (*New Scientist*, 18 June 1987, p. 63).

17   Think of the electric field generated by the power station as extending all the way along the wire as it loops from the power station around a city and back to the power station. The field stays concentrated in the wire and does not leak out because the wire is sheathed in an insulator such as PVC. This is a material that has no easily detachable electrons that can be pushed by the electric field. It therefore stops the electric field leaking out into the air and dissipating its energy.

18   For comparison, the voltage difference between a typical lightning bolt and the ground is *several hundred million volts*.

19   In fact, lightning shows the heating effect of an alternating current since a single strike consists of multiple current flows, back and forth

between the ground and a cloud, or between a cloud and a cloud.

20 Einstein, with his recognition that electric and magnetic fields are aspects of the same thing, was able to distil all electric and magnetic phenomena into a set of equations that was *even more compact* than Maxwell's.

21 Actually, a connection between electricity and magnetism and light had been suspected earlier by Michael Faraday. 'I happen to have discovered a direct relation between magnetism and light, also electricity and light, and the field it opens is so large and I think rich,' he wrote in a letter on 13 November 1845 (Georg W. A. Kahlbaum and Francis V. Darbishire, eds, *The Letters of Faraday and Schoenbein, 1836–1862*, p. 148). Among other things, Faraday had found that a magnetic field could change the plane of vibration, or polarisation, of a light wave, a phenomenon known today as Faraday rotation.

22 *The Feynman Lectures on Physics*, vol. II, pp. 1–11.

23 Ibid., pp. 1–10.

24 See Chapter 2, 'The rocket-fuelled baby: Respiration' and Chapter 5, 'Matter with curiosity: The brain'.

9 PROGRAMMABLE MATTER: COMPUTERS

1 Of course, computers have their downside. 'Imagine if every Thursday your shoes exploded if you tied them the usual way. This happens to us all the time with computers, and nobody thinks of complaining,' said Jef Raskin, an expert in human–computer interaction (Geoff Tibballs, *The Mammoth Book of Zingers, Quips, and One-Liners*).

2 With computer power increasing remorselessly, some have predicted that it will be one day possible to simulate a *universe*. In fact, philosopher Nick Bostrom thinks it is likely that such simulations have already been carried out by advanced beings an enormous number of times. If so, it is very likely that we are living in a *Matrix*-like computer-generated artificial reality! (Nick Bostrom, 'Are You Living In a Computer Simulation?', *Philosophical Quarterly*, vol. 53 (2003), pp. 243–55).

3   The first true all-purpose computer was imagined by the British engineer Charles Babbage in 1837. However, his 'analytical engine' was not built in his lifetime because of the difficulty and expense of implementing the design with mechanical cogs and wheels. Babbage worked on the project with Augusta Ada King, Countess of Lovelace and the daughter of the poet Lord Byron. She is considered the first programmer, and the computer language Ada is named in her honour.

4   There is a deep connection between Turing's discovery of uncomputability and undecidability, another great discovery in mathematics. In 1931, the Austrian logician Kurt Gödel showed that there were mathematical statements (theorems) that could never be proved either true or false. They were undecidable. Gödel's undecidability theorem – more usually known as his incompleteness theorem – is one of the most famous and shocking results in the history of mathematics. See 'God's Number', Chapter 6 of my book *The Never-Ending Days of Being Dead*.

5   Physicists have a tendency to imagine nature to be like the technological world in which they live. In the nineteenth century, in an industrial world powered by coal, for instance, they speculated that the Sun was a giant lump of coal. Today, they speculate that the Universe is a giant computer. The lesson of history suggests they are likely to be wrong, as they have been before.

6   The 'a transistor is just like a garden hose with your foot on it' image comes from *Computer Science for Fun* by Paul Curzon, Peter McOwan and Jonathan Black of Queen Mary, University of London, http://www.cs4fn.org.

7   See Chapter 8, 'Thank goodness opposites attract: Electricity'.

8   Stan Augarten, *State of the Art: A Photographic History of the Integrated Circuit*.

9   The transistor was invented by a trio of physicists at Bell Laboratories in New Jersey, USA, in 1947. For their achievement, John Bardeen, Walter Brattain and William Shockley won the 1956 Nobel Prize for Physics.

10   The integrated circuit was patented in 1959.

11 Actually, illuminated areas will remain if using a negative photo-resist; they will be dissolved if using a positive photoresist.

12 See Chapter 15, 'Magic without magic: Quantum theory'.

13 Gordon Moore, 'Cramming More Components onto Integrated Circuits', *Electronics*, vol. 38 no. 8 (19 April 1965).

14 Robert X. Cringely is actually the pen name journalist Mark Stephens and a number of other of technology writers adopted for a column in *InfoWorld*, a one-time computer newspaper.

15 Seth Lloyd, 'Ultimate physical limits to computation', *Nature*, vol. 406 no. 6799 (31 August 2000), p. 1047.

16 See Chapter 16, 'The discovery of slowness: Special relativity'.

10 THE INVENTION OF TIME TRAVEL: MONEY

1 John Médaille, 'Friends and Strangers: A Meditation on Money', *Front Porch Republic*, 20 January 2012, http://tinyurl.com/6q3pbsy.

11 THE GREAT TRANSFORMATION: CAPITALISM

1 Joseph Stiglitz, 'Inequality is holding back the recovery', in the series 'The Great Divide', *New York Times*, 19 January 2013, http://tinyurl.com/aqxj9ro.

2 The economic policies pursued in the US and much of Europe before the 2008 financial crisis involved not only deregulation but privatisation of public services and low taxes. There was a widespread belief among politicians – even Centre Left ones such as Bill Clinton and Tony Blair – that a new era in economic history had been entered and that an economy fuelled by massive debt was sustainable. In reality, what was being fuelled was a financial 'bubble', which, in common with countless other historic analogues, from the South Sea Bubble of the early eighteenth century to the dot-com bubble of the late 1990s, was bound to burst. See John Gray, *False Dawn: The Delusions of Global Capitalism*.

3 Author's telephone interview with Ja-Hoon Chang.

4 Ibid.

5   Mark Buchanan, 'Mandelbrot Beats Economics in Fathoming Markets', Bloomberg, 5 December 2011, http://tinyurl.com/75n8ecb.

6   Ronald Coase, 'The Problem of Social Cost', *Journal of Law and Economics*, October 1960.

7   Nassim Taleb, *The Black Swan: The Impact of the Highly Improbable*.

8   Mark Buchanan, 'Earthquakes and the Mind-Bending Laws of Markets', Bloomberg, 18 March 2013, http://tinyurl.com/coklmem.

9   Karl Polanyi, *The Great Transformation*.

10 David Bollier, 'Why Karl Polanyi still matters', *Commons Magazine*, 24 February 2009, http://tinyurl.com/ctktkt8.

11 Oscar Wilde, *The Picture of Dorian Gray*.

PART THREE: Earth Works

12 NO VESTIGE OF A BEGINNING: GEOLOGY

1   James Hutton, 'Theory of the Earth; or an Investigation of the Laws Observable in the Composition, Dissolution, and Restoration of Land upon the Globe', *Transactions of the Royal Society of Edinburgh*, vol. 1 (1788), pp. 209–304.

2   Most sedimentary rocks are actually created in the oceans not in lakes. However, the pioneer geologists of the eighteenth century had essentially got it right.

3   L. P. Hartley, *The Go-Between*.

4   The first transatlantic telegraph cable was laid by Isambard Kingdom Brunel's ship, the *Great Eastern*, in 1866. For his part in the feat, physicist William Thomson was knighted by Queen Victoria, finally becoming Lord Kelvin.

5   The Solar System is believed to have formed from the shrinkage under gravity of a cold cloud of interstellar gas and dust. The cloud, just like the Galaxy, was spinning. Consequently, it shrank faster between its poles than around its waist, where centrifugal force was opposing gravity. The result was a pancake-shaped cloud, with the Sun forming at the centre and leftover debris orbiting in a disc around it. Dust grains in the debris disc stuck together to make big-

ger dust grains in a runaway process that resulted in a vast number of kilometre-sized bodies. It was the collision of these planetesimals that gradually built up the planets, including the Earth. The final stages of this accretion process are recorded in the giant impact basins on the Moon.

6   See Chapter 13, 'Earth's aura: The atmosphere'.

7   Why Venus has no plate tectonics is not clear. But water is necessary to create the granite out of which continental crust is made. And Venus, being closer to the Sun than the Earth is, is believed to have lost its water to space early in its history.

8   Louis Agassiz, *Geological Sketches*.

13 EARTH'S AURA: THE ATMOSPHERE

1   Carl Sagan, 'Wonder and Skepticism', *Skeptical Enquirer*, vol. 19 no. 1, January–February 1995.

2   The scars of the Late Heavy Bombardment are evident on the face of the Moon. The impacting bodies were big enough to puncture the lunar crust, causing lava from deep inside to well up onto the surface, creating the lunar seas, or *Mare* basins. The LHB is believed to have been caused when large numbers of ice bodies from the outer Solar System were flung sunwards. The sequence of events is not completely clear. But, in one plausible scenario, Jupiter's inter-action with the Kuiper Belt, a band of icy rubble left over from the formation of the Solar System and orbiting beyond the outermost planet, Neptune, caused Jupiter to migrate outwards from the Sun. When the time taken for it to circle the Sun was exactly *half* that of Saturn, the two planets lined up regularly on the same side of the Sun, which meant that gravity of the two planets *pulled together* on other bodies in the Solar System. This caused Uranus and Neptune to change their orbits. As they moved, they stirred up the Kuiper Belt, sending large numbers of icy bodies into the inner Solar System, where they slammed into planets and their satellites, including the Earth and the Moon.

3   The best evidence that the oceans have an extraterrestrial origin

comes from observations of the water ice on Comet Hartley 2 in 2011. Hydrogen in water, H_2O, comes in two forms – the normal form, H, and a much rarer, heavier form, or isotope, known as deuterium, D. D_2O is popularly known as heavy water. The evidence from Hartley 2 is that the ratio of heavy water to water in its ice is exactly the same as in the water of Earth's oceans (Nancy Atkinson, 'Best Evidence Yet That Comets Delivered Water for Earth's Oceans', *Universe Today*, 5 October 2011, http://tinyurl.com/6zazxwj).

4   See Chapter 2, 'The rocket-fuelled baby: Respiration'.

5   The main reason the Sun heats the equatorial regions more than the polar regions is that the Sun is almost immediately above the ground at the equator while, from near the pole, it appears very low on the horizon. (At some times in the year the Sun is even *below* the horizon and out of sight.) This means that near the poles a given amount of heat is spread over a much larger area of the ground than at the equator, diluting its heating effect.

6   John Tyndall, *In Forms of Water in Clouds and Rivers, Ice and Glaciers*.

7   Venus rotates on its axis every 243 Earth days with respect to the fixed stars. Since it also orbits the Sun every 225 days, it would appear that its day is *longer* than its year. However, Venus actually rotates *backwards* compared with all other planets – except Uranus. The combination of this retrograde rotation and the planet's orbit around the Sun means that the duration of a Venusian day – from sunrise to sunset – is 117 Earth days. In other words, there are 1.92 Venusian days in a Venusian year.

8   Naively, it might be expected that heat could simply be conducted from the equator to the poles, with hot air heating the cooler air standing next to it, and that air, in turn, heating the even cooler air next to it, and so on. However, air is a bad *conductor* of heat and the atmosphere cannot conduct the necessary heat from the equator to the poles fast enough. This can happen only if there is bulk movement, or convection, of the air.

9   We do not realise that we are on a rapidly spinning planet for the

same reason we do not realise we are travelling at 900 kilometres per hour inside an airliner. Motion at constant velocity, as Galileo recognised four centuries ago, is undetectable. This is in marked contrast with *changes in our velocity*, which are very obvious. When a plane accelerates down a runway, for instance, the passengers are pinned in their seats.

10   Our common (but illusory) experience is that the ground is *not moving*. Consequently, physicists have invented a force to explain the deflection of air as it moves from the equator to the poles, *assuming that the Earth is not turning*. This Coriolis force is fictitious force but it is convenient for calculations.

11   Latitude is defined as the angular distance of any place on the surface of the globe from the equator, measured in degrees. By convention, the equator has a latitude of 0° and the poles 90°. To distinguish between places above and below the equator, a location may be said to be 34° *North* or 52° *South*.

12   The reason the deflection of north–south-moving air by the Earth's rotation is the most severe in mid-latitudes is that the deflection depends on two things: how fast the Earth's surface is rotating at the particular latitude and how fast the rotation is changing. Take the region near the poles. Although the Earth's rotation velocity is changing fast, the velocity itself is small, so the *deflection is small*. Contrast this with the tropics. Although the Earth's rotation velocity is large, it is changing only slowly, so the *deflection is again small*. But, in mid-latitudes, the Earth's rotation velocity *and* the rate at which it is changing are large. Consequently, the deflection of air as seen from the ground is significant. This is why the circulation band at mid-latitudes is the most unstable and turbulent. And why the polar and tropical bands behave like relatively well-behaved mini Hadley cells.

13   The pressure of a gas is a consequence of it being made of in-numerable atoms flying about randomly. If they drum on any obstacle, they impart on it a jittery force. This, averaged out, is the pressure. The denser the gas, the more atoms there are drumming and the greater the pressure; the hotter the gas, the faster are the

drumming atoms and, again, the higher the pressure. Because the atmosphere is unevenly heated and sloshes about, it is unavoidable that in some regions there is lower than average pressure and in others higher than average pressure.

14   Robert T. Ryan, *The Atmosphere*.

15   Edward Lorenz, 'Does the flap of a butterfly's wings in Brazil set off a tornado in Texas?', paper presented at the 139th Annual Meeting of the American Association for the Advancement of Science on 29 December 1979 (*The Essence of Chaos*, appendix 1), p. 181.

16   It might be imagined that, because the southern hemisphere receives more heat from the Sun in its summer than the northern hemisphere does in its summer, summer in the south is *hotter* than in the north. Surprisingly, it is not. The extra solar energy during the southern-hemisphere summer principally heats the top of the atmosphere but this is not translated to the surface below. The temperature at any place on the surface depends not only on solar heating but on a complex mix of factors, including the amount of cloud cover, and the speed at which the air and ocean currents transport heat to and from the location.

17   It is a characteristic of a spinning body such as the Earth that it doggedly maintains its spin direction in space (this is why we get *regular* seasons). However, over long periods of time, the combined gravitational pulls of the Sun, Moon and planets tug on the Earth and make it wobble like a top. This precession causes the planet's spin axis – still maintained at 23.5° to the vertical – to rotate about the vertical *once every 26,000 years*. Currently, the star roughly above the North Pole – that is, in the direction of the Earth's spin axis if it were extended into space – is Polaris. Because of the effect of precession, however, Polaris was not the Pole Star when the Pyramids were built 5,000 years ago. And it will not be the Pole Star 5,000 years hence. Only in 26,000 years' time will it return to where it is today.

18   The magnetic field of the Sun, like that of the Earth, is generated by electrically charged material circulating deep inside. Because the

Sun rotates faster at the equator than at its poles, the magnetic field emerging from the Sun becomes twisted up, eventually releasing its pent-up energy in flares. For some reason, which is not clear, the cycle of build-up and release of energy takes about twenty-two years. This is the solar cycle. During the active part of this solar cycle, the surface of the Sun is covered by many sunspots and violent flares.

19 Ultraviolet is a type of light emitted by matter at extremely high temperature. On the Sun, it is produced by flares – fountains of matter hurled into space by pent-up magnetic energy. These can easily attain temperatures of 10 to 20 million °C.

20 See Chapter 12, 'No vestige of a beginning: Geology'.

21 The Sun shines by fusing the cores, or nuclei, of the lightest element, hydrogen, into the nuclei of the second lightest, helium. The by-product of this nuclear reaction is sunlight. Helium, being heavier than hydrogen, sinks to the core of the Sun, where gravity squeezes it tightly. As anyone who has squeezed the air in a bicycle pump knows, when a gas is squeezed it gets hotter. Consequently, as the Sun turns hydrogen into helium, its core – and consequently the whole Sun – gets hotter.

22 A big puzzle is why did the Earth not freeze solid if the Sun was 30 per cent less luminous at the planet's birth? One possible solution to the faint young sun paradox is that the newborn Earth was shrouded in a thick enough atmosphere of greenhouse gases and that their warming effect prevented the planet plunging into an interminable ice age.

23 Many astronomy books say the Earth will be swallowed by the Sun which, as a red giant, will balloon out almost to the orbit of Mars. However, a team led by Juliana Sackmann of the California Institute of Technology in Pasadena has pointed out that, although the Sun will get to the Earth's orbit, when it does, the Earth *will not be there*. The reason is that red giants lose material at a terrific rate via their stellar winds. A less massive Sun will have weaker gravity with which to hold onto the Earth, so the Earth will gradually move away. By the time the Sun reaches the Earth's current orbit, it will have

only 60 per cent of its present mass and the Earth will be 70 per cent further away, so the planet will probably escape being gobbled. A team led by Mario Livio of the Space Telescope Science Institute in Baltimore, however, points out there is a competing effect. The Earth raises a tidal bulge in the Sun, which it will try to drag around as it orbits. As a consequence, the Earth will spin up the envelope of the Sun while it slows and moves inward. The rate at which the Earth is sapped of orbital energy depends crucially on how viscous is the stuff of the Sun's envelope, which nobody knows well. Currently, therefore, it is not possible to tell which of the two effects will win and whether or not the Earth will be gobbled.

PART FOUR: Deep Workings

14 WE ARE ALL STEAM ENGINES: THERMODYNAMICS

1   It is not quite true that the Earth gains no net energy from the Sun. It gains a little. For instance, the level of the greenhouse gas carbon dioxide, which traps heat in the atmosphere, is rising. This is causing global warming. Trees also sequester some solar energy which, if they fall and are buried, may become coal in many millions of years' time. Coal is trapped sunlight and, when we burn it, we free yesterday's sunlight.

2   Peter Atkins, *Four Laws that Drive the Universe*.

3   The difference between heat and temperature – the *degree of hotness* of a body – is illustrated by a match and a central-heating radiator. A match contains very little heat but has a temperature high enough to burn you. A radiator contains a lot of heat but has a temperature low enough for you to lean safely against it.

4   Atoms tend to combine to form molecules under the influence of their mutual electromagnetic force. A molecule of steam, for instance, consists of two hydrogen atoms glued to one oxygen atom (H_2O).

5   Actually, the reason the molecules of steam lose speed is subtle. If the piston was not moving, they would bounce off it like perfect

rubber balls, losing no speed. However, the piston is moving *away* from the molecules. This means that, when a molecule bounces off the piston, its speed relative to it is less than it would have been relative to a stationary piston.

6   Conservation laws are nothing more than manifestations of deep symmetries. These are properties of the world that stay the same under a particular transformation. For instance, the law of conservation of energy is a consequence of time-translation symmetry, the fact an experiment done today or next week will produce the same result. This idea that symmetry underpins the laws of physics was discovered by the German mathematician, Emmy Noether, in 1918, and is one of the most powerful ideas in all of science. See Chapter 20, 'Rules of the game: The laws of physics'.

7   The efficiency of a steam engine that uses steam at a temperature of T_h and discharges waste heat to its surroundings at T_c is $1 - T_c/T_h$ (with the temperature expressed in Kelvin; see n. 10 below). The formula was discovered by the nineteenth-century French engineer, Sadi Carnot. It shows, for instance, that, if an engine uses steam at 373 Kelvin and discharges waste heat to its surroundings at 300 Kelvin, it can turn only about 20 per cent of the energy of the steam into useful work.

8   We can ignore the piston because temperature and entropy describe only disordered microscopic motion. The piston exemplifies ordered bulk motion.

9   Arthur Eddington, *The Nature of the Physical World*.

10   For temperature, physicists tend to use the Kelvin scale. This assigns 0 Kelvin to the temperature at which microscopic motion becomes so sluggish that it actually stops altogether. Since on the Celsius scale this is -273 °C, the freezing point of water, 0 °C, is equivalent to 273 Kelvin. This makes the average surface temperature of the Earth about 300 Kelvin.

11   Although the light that falls on the Earth from the Sun is predominantly visible light – characteristic of a body glowing at 5,778 Kelvin – the light the Earth radiates into space is invisible-to-the-naked-eye far infrared – characteristic of a body at about 300 Kelvin.

12   For every 5,778 Kelvin photons the Earth receives from the Sun it radiates back into space about twenty 300 Kelvin photons. Every photon, it turns out, has about the same entropy. So the Earth exports to the Universe about twenty times the entropy it receives from the Sun. All this extra disorder is the price paid by the Universe for all the wonderful things that go on on Earth.

13   Since all activity in the Universe is driven by the temperature distance between the stars and empty space, the obvious question is: what created that temperature difference? The answer is gravity. Shortly after the big bang, the matter of the Universe was spread out uniformly at uniform temperature. But regions that were slightly denser than average had slightly stronger gravity and began to drag more matter towards them. The ultimate result was to squeeze matter into dense clumps – and when matter is squeezed it becomes hot. Gravity, then, changed a tepid, uniform Universe into a Universe full of clumpy hot things – stars.

14   Entropy is related to our *lack of information* about a system – in other words, to our ignorance of it. If the energy is in disordered steam, for instance, it is impossible to know which of the countless molecules have the energy of motion. High entropy is therefore the same as having a high level of ignorance. However, when a piston is moving, it is obvious where the energy of motion is – with the whole piston. Low entropy is therefore synonymous with having a low level of ignorance.

15   Howard Resnikoff, *The Illusion of Reality*.

16   Actually, since 1998, we have known that the expansion of the Universe is speeding up, driven by the repulsive gravity of mysterious dark energy. The fate of matter may therefore be to be diluted out of existence by the breakneck expansion. It could be *even more boring* than anyone suspected.

15 MAGIC WITHOUT MAGIC: QUANTUM THEORY

1   See Chapter 18, 'The roar of things extremely small: Atoms'.
2   Werner Heisenberg, *Physics and Philosophy*.
3   Werner Heisenberg, *Quantum Theory*.

4   There is an interesting parallel here with space–time. Space–time, being 4-dimensional, is ungraspable by 3-dimensional creatures such as us. Instead, we experience merely facets of space–time – space and time (see Chapter 16, 'The discovery of slowness: Special relativity'). In the same way, we see only the particle-like and wave-like facets of light.

5   If the Universe were not fundamentally unpredictable, there would not be a Universe – or at least a Universe of the complexity necessary for us to be here. The reason is that, according to the standard picture of cosmology, known as inflation, the Universe started out so ultra-tiny that it contained hardly any information. Today, it contains a truly vast amount – just imagine how much would be needed to describe the type and location of every atom in the Universe. The puzzle of where all the information came from is explained by quantum theory since randomness is synonymous with information. Every quantum event since the big bang, such as the decay of a radioactive atom, has happened randomly, injecting information/complexity into the Universe. When Einstein said, 'God does not play dice with the Universe', he could not have been more wrong. If God had not played dice, there would be no Universe – certainly no Universe with anything interesting going on in it. See 'Random Reality', Chapter 10 of my book *We Need to Talk about Kelvin*.

6   Technically, the probability of finding the atom at a particular location is the square of the amplitude of the quantum wave, or wave function, at that location.

7   'I bet that was fun for the rest of the Thomson family at get-togethers. "It is. It isn't! It is . . . !"' said @Katharine_T29m, one of my Twitter followers.

8   The incredible thing is that, even if Davisson, Germer and Thomson had fired their electrons at their crystal *one at a time*, *with an hour gap between each one*, over time they would have observed exactly the same pattern: there would have been directions in which electrons are *seen* alternating with directions in which they are *never seen*. So it is not interference between the quantum waves of *different*

electrons that creates the pattern. It is interference between quantum waves of a single electron. Each electron is in a superposition corresponding to it going in all directions at once and it is the individual waves of this superposition that interfere with each other. Quantum theory is truly mind-bending.

9   Technically, spin ½ means that an electron has a spin of ½ × $(h/2^*\pi)$, where h is Planck's constant.

10   If history had run itself differently, not only would the particle with the smallest spin have been assigned a spin of 1 unit, the particles with the smallest electric charge would have been assigned a charge of 1 unit. Instead, the electron has ended up with a spin of ½, and the quarks charges of magnitude ⅓ and ⅔.

11   See Chapter 8, 'Thank goodness opposites attract: Electricity'.

12   See Chapter 16, 'The discovery of slowness: 'Special relativity'.

13   See Chapter 8, 'Thank goodness opposites attract: Electricity'.

14   The wavelength of a particle, as Louis de Broglie guessed in 1923, is inversely proportional to its momentum. To be specific, it is $(h/2^*\pi)/p$, where p is the momentum.

15   The resistance of an electron wave to being squashed gives rise to a force known as electron degeneracy pressure. In 5 billlion years' time, when the Sun has exhausted its heat supply, gravity will gain the upper hand and shrink it down to the size of the Earth. The force that will prevent it shrinking any more than this will be electron degeneracy pressure – the resistance of electron waves to being squeezed. From the particle point of view – which is more complicated than the wave point of view – the force is said to be due to the Heisenberg Uncertainty Principle. This merely says that the smaller the volume in which a particle is confined, the greater its momentum. Think of a bee that buzzes about more angrily the smaller the box in which it is confined.

16   See Chapter 18, 'The roar of things extremely small: Atoms'.

17   'I'm more of a glass is 0.000000000000001% full kinda person myself,' said @MrDFJBaileyEsq, one of my Twitter followers.

18   A planet cannot quite orbit anywhere in the Solar System. The space between the orbits of Mars and the giant planet Jupiter, for instance,

is populated only by chunks of rocky rubble, or asteroids. A fully fledged planet was prevented from forming here by the disruptive effect of Jupiter's powerful gravity.

19 The details of how electron spin, waviness and indistinguishability spawn the Pauli Exclusion Principle are described in 'No More than Two Peas in a Pod at a Time', Chapter 3 of my book *We Need to Talk About Kelvin*.

16 THE DISCOVERY OF SLOWNESS: SPECIAL RELATIVITY

1 Although relativity predicts that someone moving relative to you should appear to shrink in the direction of their motion, this is not exactly what you would see. There is another effect at play. Light takes longer to reach you from more distant parts of the person than from nearer parts. This causes them to appear to rotate. So, if their face is pointing towards you, you will see some of the back of their head. This peculiar effect is known as relativistic aberration, or relativistic beaming.

2 The disintegration, or decay, of muons is an unpredictable, random process. However, physicists talk of their half-life. After a period of one half-life, half the muons are left; after two half-lives, half as many again – that is, a quarter; after three half-lives, one-eighth, and so on.

3 This scenario is not difficult to imagine. Think of two fireworks that appear to go off at the same time from the point of view of someone standing midway between them. Switch to the point of view of someone else who sees one firework behind the other. The light from the most distant explosion will arrive later at their location, so they will see the two events at different times.

4 See Chapter 18, 'The roar of things extremely small: Atoms'.

5 According to an idea proposed in 1964 by English physicist Peter Higgs and five others, the mass of fundamental particles such as the electron is not intrinsic but *extrinsic*. It is endowed on them by their interaction with the Higgs field, which pervades all of space. The field is like an invisible cosmic treacle that impedes the passage of

subatomic particles. Resistance to motion is what we think of as mass. If you push a loaded fridge, it resists. In the Higgs picture, this is because you are pushing it through the cosmic treacle. The Higgs particle is the quantum of the Higgs field, just as the electron is the quantum of the electric field.

6 Technically, the space–time interval that is the same for all observers is $\sqrt{(x^2 + y^2 + z^2 - c^2 t^2)}$, where x, y, z are the space interval between events.

17 THE SOUND OF GRAVITY: GENERAL RELATIVITY

1 See Chapter 16, 'The discovery of slowness: Special relativity'.

2 If something accelerates at 9.8 metres per second per second it merely means that, every second, it gets faster by 9.8 metres per second.

3 James Chin-Wen Chou et al., 'Optical Clocks and Relativity', *Science*, 24 September 2010, vol. 329, p. 1630.

4 A black hole is a region of space–time where gravity is so strong that nothing, not even light, can escape. Such a region is left when a very massive star reaches the end of its life and its core shrinks catastrophically under its own gravity. See Chapter 22, 'Masters of the Universe: Black holes'.

5 Photons have no intrinsic, or rest, mass. Their effective mass is entirely due to their energy, or momentum.

6 If gravity is not quite the same as acceleration, it might appear to undermine the whole basis of general relativity. However, gravity and acceleration are always indistinguishable *locally* – that is, in a small enough region of space. And this, it turns out, is enough of a foundation on which to build Einstein's theory of gravity.

7 Total eclipses of the Sun by the Moon are possible because of a very fortunate coincidence. Although the Sun is about 400 times further away than the Moon, it is also about 400 times bigger. Consequently, the Sun and the Moon have the same apparent size in the sky. The Moon is moving away from the Earth at about 4 centimetres a year. This means that total eclipses will not be visible in 100 million years'

time. Nor were they visible at the time of the dinosaurs 100 million years ago.

8   Einstein's fields equations (1915) are: $G^{mn} = -(8pG/c^2)T^{mn}$. In words, they say that the warpage, or geometry, of space–time (G^{mn}) is generated by matter and energy (T^{mn}). Each superscript represents 1 of the 4 coordinates of space–time so there are actually $4 \times 4 = 16$ equations. But, since some are repeated, this reduces to 10. This is still 10 times as many as are required for Newton's law of gravity.

9   According to Newton, gravity is a force of attraction between *all* bodies. So not only is there a force between the Sun and the Earth, there is a force between you and a person standing next to you, between you and the coins in your pocket. The force is extremely weak but grows with mass, which is why people passing each other in the street are not snapped together, whereas the Earth is trapped by the Sun. The force is mutual – in other words, the Earth exerts the same gravitational force on you as you do on the Earth. The reason you are affected by the Earth more than the Earth is affected by you is simply that you are smaller and easier to move. ('Is that why I am attracted to big women but big women are not attracted to me?' asked the English comedy writer Andy Hamilton on the pilot of the BBC4 comedy-science series *It's Only a Theory*. He was highlighting a profound truth!)

10 Strictly speaking, a body moving under the influence of the inverse-square-force of another body traces out a conic section – an ellipse, parabola or hyperbola. The path is an ellipse if the body has insufficient energy to escape its gravitational entrapment; a hyperbola if it has; and a parabola if the body is teetering on the knife edge between being trapped and escaping to infinity.

11 A spectrum is formed when light is fanned out, or separated, into its constituent colours. In the past half a century, our vision, sensitive to a mere handful of rainbow hues, has been artificially enhanced to reveal a billion new colours arrayed along the electromagnetic spectrum – from gamma rays to radio waves. See Chapter 8, 'Thank goodness opposites attract: Electricity'.

12   A neutron star is the super-dense relic of a supernova explosion. Paradoxically, when a massive star at the end of its life blows off its outer layers, its core *implodes*. A neutron star contains about the mass of the Sun compressed into only the volume of Mount Everest. Consequently, a sugar cube of neutron-star stuff weighs about as much as the entire human race. See Chapter 18, 'The roar of things extremely small: Atoms'.

13   See Chapter 22, 'Masters of the Universe: Black holes'.

14   A handful of neutrinos have also been detected from beyond the Sun. So too have cosmic rays, atomic nuclei possibly sprayed into space by supernova explosions. But, essentially, all we know about the Universe comes via light we pick up with our telescopes.

18 THE ROAR OF THINGS EXTREMELY SMALL: ATOMS

1   Richard Feynman, *The Feynman Lectures on Physics*, vol. 1.

2   The idea that the world is, at its root, simple is a powerful one. It is the unspoken act of faith that has driven physics since Newton. No one knows why it is true. However, it has undoubtedly been successful in guiding us to uncover ever deeper and simpler laws of nature.

3   Individual atoms were first seen *directly* only in 1980. Gerd Binnig and Heinrich Rohrer of IBM in Zurich, Switzerland, invented the Scanning Tunnelling Microscope. The STM senses the up-and-down motion of a super-fine needle as it is dragged across the surface of a material. Think of a blind person building up a picture of someone by dragging their finger across their face. Using their STM, Binnig and Rohrer were able to 'see' the atomic landscape. Atoms looked like tiny footballs, like oranges stacked in boxes, just as Democritus had imagined them more than 2,000 years before. For their invention of the STM, Binnig and Rohrer won the 1986 Nobel Prize for Physics.

4   See Chapter 14, 'We are all steam engines: Thermodynamics'.

5   James Clerk Maxwell, 'On the Motions and Collisions of Perfectly Elastic Spheres', *Philosophical Magazine*, January and July 1860.

6 Tom Stoppard, *Hapgood*.

7 Although a single neutrino is hardly ever stopped by an atom, these elusive particles can be detected by putting in their path a *lot of atoms*. The SuperKamiokande detector, deep in a mountain in Japan, is a 14-storey 'baked-bean can' filled with 50,000 tonnes of ultrapure water. Occasionally, a neutrino interacts with a proton in a water molecule. The subatomic shrapnel flies outwards through the water and creates the light equivalent of a sonic boom. This blue, Cherenkov, light – characteristic of ponds holding spent nuclear fuel – is picked up by light detectors, which cover the interior of the baked-bean can. SuperKamiokande has produced one of the most amazing images in the history of science. It is a picture of the Sun, taken at night, not looking up at the sky but down through 12,760 kilometres of rock to the other side of the Earth, not with light but with neutrinos. http://tinyurl.com/ao4wdny.

8 While neutrinos take just 2 seconds to emerge from the Sun and a further 8½ minutes to travel across space to the Earth, sunlight takes about 30,000 years to get out of the Sun. Consequently, today's sunlight is about 30,000 years old. It was made at the height of the last ice age.

9 See Chapter 16, 'The discovery of slowness: Special relativity'.

10 See Chapter 15, 'Magic without magic: Quantum theory'.

11 Technically, the Pauli Exclusion Principle is a consequence of the fact that particles such as electrons are (1) indistinguishable, (2) behave like waves, and (3) behave like fermions, which technically means they have half-integer spin. See Chapter 15, 'Magic without magic: Quantum theory'.

12 It is possible to have more neutrinos if they are of a type known as sterile. The normal neutrinos, although antisocial, *do* interact with normal matter very occasionally via nature's weak nuclear force. Sterile neutrinos would not even do this. Their sole interaction with normal matter would be via the gravitational force.

13 See Chapter 16, 'The discovery of slowness: Special relativity'.

14 See Chapter 17, 'The sound of gravity: General relativity'.

15 Technically, fermions have half-integer quantum spin and bosons

have integer spin. This leads to fermions obeying the Pauli Exclusion Principle – which means they are very antisocial – and bosons ignoring the Pauli Principle – which makes them very gregarious (see 'No More than Two Peas in a Pod at a Time', Chapter 3 of my book *We Need to Talk About Kelvin*).

16 See Chapter 21, 'The day without a yesterday: Cosmology'.

17 Murray Gill-Mann, 'What Is Complexity?', *Complexity*, vol. 1 no. 1, 1995.

18 See 'Random Reality', Chapter 10 of my book *We Need to Talk About Kelvin*.

19 Forest Ray Moulton (ed.), *The Cell and the Protoplasm*, p. 18.

19 NO TIME LIKE THE PRESENT: TIME

1 Douglas Adams, *The Hitch Hiker's Guide to the Galaxy*.

2 The 'surface of last scattering' marks the point at which the fireball of the big bang had cooled sufficiently for nuclei and electrons to combine to make the first atoms. Whereas free electrons are good at scattering, or redirecting, light, electrons in atoms are not. Consequently, before the epoch of last scattering, the Universe was a fog, impenetrable to light. After, light was able to travel unhindered in straight lines, and the Universe became transparent. Today, we see the light from this epoch as the cosmic background radiation. See Chapter 21, 'The day without a yesterday: Cosmology'.

3 If the Universe is a place where space is turned into time, by contrast, 'museums are places where Time is transformed into Space' (Orhan Pamuk, *The Museum of Innocence*).

4 See Chapter 16, 'The discovery of slowness: Special relativity'.

5 See Chapter 17, 'The sound of gravity: General relativity'.

6 Charles Misner, Kip Thorne and John Wheeler, *Gravitation*, p. 937.

7 See Chapter 14, 'We are all steam engines: Thermodynamics'.

8 The second law of thermodynamics, for all its subtlety, is virtually a tautology. As American physicist Larry Schulman of Clarkson University in New York, points out, it really says only: 'More likely things are more likely to happen.'

9   The number of ways the components of a body can be rearranged and still make the body – technically, the 'number of microstates corresponding to a macrostate' – is dubbed W in physics. The entropy, S, is then given by: $S = k \log W$, where k is known as Boltzmann's constant. This is the definitive statement of the second law of thermodynamics. It is one of the most beautiful and powerful equations in all of physics and it is inscribed on the headstone of Boltzmann's grave in Vienna.

10   A highly ordered state synonymous with a *highly unlikely state.* Consequently, this has caused much unease among physicists. However, Larry Schulman says the key to why the early Universe was in a highly ordered state is the 'epoch of last scattering'. At this time, about 379,000 years after the birth of the Universe, the big-bang fireball had cooled enough for atomic nuclei and electrons to combine into the first atoms. Crucially, free electrons interact strongly with photons whereas electrons trapped in atoms do not. Since there were about 10 billion photons of the big-bang fireball for every electron, this meant that, before the Universe was 379,000 years old, matter was blasted apart before gravity could pull it together. Afterwards, however, gravity could no longer be thwarted. And gravity is the key. The matter of the cooling fireball was spread extremely smoothly throughout space. But, although this is the most likely, disordered, state in the absence of gravity, in the presence of gravity, it is actually a very unlikely, ordered, state (the most likely state for matter in the presence of gravity is clumped, as can be seen from the galaxies and stars in today's Universe). So, even though the distribution of matter in the Universe did not change at the epoch of last scattering, the 'switching on' of gravity was responsible for the Universe suddenly finding itself in an extremely unlikely, ordered, state. (See L. S. Schulman, 'Source of the Observed Thermodynamic Arrow', *Journal of Physics: Conference Series*, vol. 174 no. 1 (2009), 12,022.) A similar argument to Schulman's has been proposed by the British physicist Roger Penrose.

11   James Hartle, 'The Physics of Now', *American Journal of Physics*,

vol. 73, issue 2 (February 2005), p. 101; http://arxiv.org/abs/gr-qc/0403001.

20 RULES OF THE GAME: THE LAWS OF PHYSICS

1   Philip Anderson, 'More Is Different', Science, vol. 177 no. 4047 (4 August 1972).

2   'Laws of Physics for Cats', http://www.funny2.com/catlaws.htm.

3   Actually, Newton's law of gravity turns out to be true only when gravity is relatively weak, which is in most normal circumstances. The theory that describes the behaviour of gravity, both weak *and* strong, is Einstein's general theory of relativity. See Chapter 17, 'The sound of gravity: General relativity'.

4   One of the central characteristics of a scientific theory, as encapsulated in a scientific law, is that *you get more out than you put in*. Pseudoscientific explanations all fall at this hurdle. To get out a lot, it is generally necessary to put in a lot. In fact, in the case of Creationism, in order to explain the Universe, it is necessary to postulate something even more complicated than the Universe – namely, God. This amounts to putting in more than you get out, the very opposite of science.

5   One controversial explanation for why mathematics is such a perfect metaphor for physics is that mathematics *is* physics. The Swedish-American physicist Max Tegmark has taken the increasingly popular idea that we live in one universe within a vast ensemble of other universes, or a 'multiverse', and run with it. He says every discrete piece of mathematics is implemented in a universe. In other words there is a universe with only flat-paper geometry, another with Boolean logic, and so on. But most of these universes are dead. Only in universes with mathematics/physics complicated enough to generate intelligence will intelligence arise. We live in such a universe, says Tegmark. After all, how could we not?! (Max Tegmark, 'Is the "Theory of Everything" Merely the Ultimate Ensemble Theory?', *Annals of Physics*, vol. 270, issue 1 (20 November 1998), pp. 1–51.)

6   Neil deGrasse Tyson, *Death by Black Hole: And Other Cosmic Quandaries*.

7   Alan Sokal, 'A Physicist Experiments with Cultural Studies', *Lingua Franca*, May/June 1996; http://tinyurl.com/mvow.

8   Fundamental physics is the search for laws that are not dependent on our viewpoint – for instance, on how fast we are moving or how strong is the gravity we are experiencing – laws, that is, that we can all agree on. In relativity, such observer-independent laws are called 'covariant'.

9   Actually, in 2012, I saw a four-armed starfish near Broome in Western Australia.

PART FIVE: The Cosmic Connection

21 THE DAY WITHOUT A YESTERDAY: COSMOLOGY

1   The first galaxies to form were actually relatively small. But, over the past 10 billion years or so, they have repeatedly merged and cannibalised each other, growing ever bigger until finally creating the galaxies we see around us today.

2   John Haines, 'Little Cosmic Dust Poem' (1983); http://tinyurl.com/crwo3y4.

3   Strictly speaking, the cosmic background radiation is brightest at a far-infrared wavelength of about 1 millimetre. Historically, however, it was first spotted at the easier-to-detect microwave wavelength of a few centimetres.

4   Strictly speaking, it is necessary to be at high altitude or in space to see the Universe glowing with the relic heat of the big bang. This is because water vapour in the atmosphere strongly absorbs the far infrared of the cosmic background radiation. At altitude, this water vapour is frozen out.

5   At one time there was a rival of the big bang. In 1948, Fred Hoyle, Hermann Bondi and Thomas Gold proposed that, although the Universe is expanding, new material continually pops into existence out of nothing to make new galaxies, so the Universe never gets more

dilute but always looks the same. The steady state was dealt a killer blow by the discovery that the distant, and therefore ancient Universe, looks very different from today, and by the discovery of the cosmic background radiation in 1965.

6   When hydrogen nuclei approach each other close enough, they come under the influence of the powerful nuclear force. Like pieces of shrapnel in an explosion in reverse, they begin to fall towards each other. Faster and faster they fall until, finally, they collide. By the time this happens, however, they have acquired a tremendous energy of motion, which they must somehow get rid off if they are to stick together rather than rebound outwards. The surplus energy might be lost in the form of a high-energy particle or gamma ray. The details are unimportant. The key thing is that the formation of a nucleus of helium out of hydrogen nuclei is accompanied by a loss of a large amount of energy. This is the ultimate origin of sunlight. See my *The Magic Furnace*.

7   Absolute zero, equivalent to -273.15 °C, is the lowest possible temperature. Classical, or pre-quantum, physics predicts that, as the temperature falls, the jiggling of atoms gets ever more sluggish. At absolute zero, it stops altogether.

8   The cosmic background radiation broke free of matter about 379,000 years after the birth of the Universe. It had existed before but its photons could barely travel across space without being redirected, or scattered, by free electrons. About 379,000 years old, the Universe had cooled sufficiently for electrons to combine with atomic nuclei to make the first atoms. Without free electrons to hinder them, the photons of the fireball were suddenly free to travel across space unhindered. We detect them today as the cosmic background radiation. They have come directly to us from this epoch of last scattering.

9   The speed of light is the cosmic speed limit only in Einstein's special theory of relativity of 1905. In his general theory of relativity of 1915, space can expand at any speed it likes. Evidence for this faster-than-light expansion comes from the size of the observable Universe. Although the Universe has existed for only 13.8 billion years, it is 84 billion light years across.

10   According to quantum theory, the vacuum is not empty. Far from it. Whereas in the everyday world the law of conservation of energy forbids energy being created from nothing, in the subatomic world, nature overlooks this edict. Energy can be conjured out of nothing, *as long as it is paid back quickly*. Think of a teenager who gets away with borrowing his dad's car overnight as long he returns it to the garage the following morning before his dad notices its absence. In the same way, nature turns a blind eye to energy being conjured out of nothing as long as it is for only an ultra-short time. Consequently, the quantum vacuum, far from being empty, seethes with restless energy.

11   According to Einstein's equations of gravity, the source of gravity is $u + 3P$, where u is energy density and P is the pressure. Usually, the second term is ignored because, in normal circumstances, the pressure of matter – due to its microscopic components buzzing about – is negligible compared with the energy density of matter. But it is always possible that there exists some hitherto unknown type of material in which the pressure is not negligible. And, if the pressure P is negative and less than $-u/3$, this reverses the sign of $u + 3P$, making gravity *repulsive* – that is, gravity *blows rather than sucks*. This is the case with the inflationary, false, vacuum. Incidentally, negative pressure means that, instead of pushing outwards, the vacuum is trying to shrink everywhere. Yet, bizarrely, it has repulsive gravity, and inflates. The reason for this is that the pressure has no direct effect. Every chunk of shrinking vacuum is surrounded by other chunks of shrinking vacuum so, overall, the negative pressure cancels out. Instead, the negative pressure has an indirect effect entirely through Einstein's equations, which endow it with repulsive gravity.

12   This is typical of anything quantum. Its behaviour – for instance, whether it decays – is totally random, totally unpredictable. The Universe was a quantum object in its first split second of existence because it was *smaller than an atom*.

13   To be fair, nobody has yet managed to unite quantum theory with Einstein's theory of gravity. A quantum theory of gravity, for all

we know, may predict the *exact* energy density observed for the dark energy. Uniting quantum theory – a theory of the very small – and the general theory of relativity – a theory of the very big – is essential to understanding the birth of the Universe. After all, at that time, something very big was also very small.

14   The clumping together of matter to form galaxies could begin only when the fireball had cooled enough for electrons to combine with nuclei to form the Universe's first atoms. The reason is that free electrons interact strongly with, or scatter, photons, and there were about 10 billion for every electron in the big-bang fireball. They blasted apart any matter trying to clump together under gravity. Once electrons were mopped up by atoms, however, gravity gained control of the Universe. The time when galaxies began to form, about 379,000 years after the birth of the Universe, is known as the epoch of last scattering. Priceless information about this period is imprinted on the cosmic background radiation.

15   The evidence for dark matter also comes from within galaxies. The stars in the outer regions of spiral galaxies such as our Milky Way, for instance, are orbiting *too fast*. Like children on a speeded-up roundabout, they should fly off into intergalactic space. The reason they do not, astronomers maintain, is that they are in the grip of the gravity of a huge mass of dark matter. This dark matter, which greatly outweighs the visible stars, is believed to form a spherical halo in which is embedded the flattened disc of the spiral galaxy.

16   Earlier, it was mentioned that a crucial piece of evidence for the big bang is that 25 per cent of the mass of the Universe is helium. That is 25 per cent of the *ordinary matter*.

17   See 'The Holes in the Sky', Chapter 6 of my book *The Universe Next Door*.

18   Cosmological parameters such as the age and expansion rate of the Universe were once very badly known. Everything changed with the launch in 2001 of NASA's Wilkinson Microwave Anisotropy Probe to observe the afterglow of the big bang. It ushered in the age of precision cosmology.

19   Apologies for using the image of a bubble in two different contexts.

Each bubble that forms in the inflationary vacuum actually contains an *infinite* number of big-bang regions (smaller bubbles), each like our observable Universe. If you are wondering how something can be bounded yet infinite, it is because the inflationary vacuum is expanding so incredibly fast that, to observers inside the bubble, the boundary is *unreachable*. Effectively, therefore, the bubble is infinite.

20   Arthur C. Clarke, 'The Wall of Darkness', *The Other Side of the Sky*.

22 MASTERS OF THE UNIVERSE: BLACK HOLES

1   See Chapter 17, 'The discovery of slowness: General relativity'.

2   Kenneth Ford and John Wheeler, *Geons, Black Holes and Quantum Foam*.

3   Although John Wheeler is often credited with coining the term 'black hole', he in fact merely popularised it. 'In the fall of 1967, [I was invited] to a conference . . . on pulsars,' he wrote in *Geons, Black Holes, and Quantum Foam*. 'In my talk, I argued that we should consider the possibility that the center of a pulsar is a gravitationally completely collapsed object. I remarked that one couldn't keep saying "gravitationally completely collapsed object" over and over. One needed a shorter descriptive phrase. "How about black hole?" asked someone in the audience. I had been searching for the right term for months, mulling it over in bed, in the bathtub, in my car, whenever I had quiet moments. Suddenly this name seemed exactly right. When I gave a more formal Sigma Xi-Phi Beta Kappa lecture . . . on December 29, 1967, I used the term, and then included it in the written version of the lecture published in the spring of 1968. (As it turned out, a pulsar is powered by "merely" a neutron star, not a black hole.)'

4   David Elbaz et al., 'Quasar Induced Galaxy Formation: A New Paradigm?', *Astronomy and Astrophysics*, vol. 507 no. 3 (1 December 2009), pp. 1359–74; http://arxiv.org/abs/0907.2923.

5   Physics is a recipe for predicting the future – in classical physics, one future with 100 per cent certainty; in quantum physics, the probabilities of a range different futures. Newton's law of gravity, for

instance, allows physicists to predict the location of the Moon tomorrow from its location today. In a sense, then, the location of the Moon tomorrow is *contained in its location today*. No new information is introduced. Information in physics is neither created nor destroyed. Ironically, Stephen Hawking actually believed that Hawking radiation showed that black holes were exceptional objects that bucked nature's trend. 'I used to think information was destroyed in a black hole,' he said. 'This was my biggest blunder, or at least my biggest blunder in science.'

6   See Chapter 14, 'We are all steam engines: Thermodynamics'.

BIBLIOGRAPHY

Adams, Douglas, *The Hitch Hiker's Guide to the Galaxy* (Macmillan, 2009)

Agassiz, Louis, *Geological Sketches* (Ticknor and Fields, 1866)

Anderson, Philip, 'More Is Different', *Science*, vol. 177 no. 4047 (4 August 1972)

Atkins, Peter, *Four Laws that Drive the Universe* (Oxford University Press, 2007)

Atkinson, Nancy, 'Best Evidence Yet That Comets Delivered Water for Earth's Oceans', *Universe Today*, 5 October 2011, http://tinyurl.com/6zazxwj

Augarten, Stan, *State of the Art: A Photographic History of the Integrated Circuit* (Houghton Mifflin, 1983)

Bais, Sander, *Very Special Relativity* (Harvard University Press, 2007)

Barash, David, *Homo Mysterious: Evolutionary Puzzles of Human Nature* (Oxford University Press, 2012)

Begley, Sharon, 'In Our Messy, Reptilian Brains', *Newsweek*, 9 April 2007

Berners-Lee, Tim, *Weaving the Web: The Past, Present and Future of the World Wide Web by its Inventor* (Orion Business, 1999)

Bierce, Ambrose, *The Devil's Dictionary* (1911)

Bollier, David, 'Why Karl Polanyi still matters', *Commons Magazine*, 24 February 2009, http://tinyurl.com/ctktkt8

Bostrom, Nick, 'Are You Living In a Computer Simulation?', *Philosophical Quarterly*, vol. 53 (2003), pp. 243–55

Buchanan, Mark, 'Mandelbrot Beats Economics in Fathoming Markets', Bloomberg, 5 December 2011, http://tinyurl.com/75n8ecb

—— *Forecast: What Physics, Meteorology and the Natural Sciences Can Teach Us about Economics* (Bloomsbury Publishing, 2013)

—— 'Earthquakes and the Mind-Bending Laws of Markets', Bloomberg, 18 March 2013, http://tinyurl.com/coklmem

Butler, Samuel, *Life and Habit* (1878)

Chang, Ha-Joon, *23 Things They Don't Tell You About Capitalism* (Penguin, 2010)

Chin-Wen Chou, James, et al., 'Optical Clocks and Relativity', *Science*, 24 September 2010, vol. 329, p. 1630

Chown, Marcus, *The Magic Furnace* (Vintage, 2000)

— *The Universe Next Door* (Headline, 2002)

— *The Never-Ending Days of Being Dead* (Faber and Faber, 2007)

— *Quantum Theory Cannot Hurt You* (Faber and Faber, 2008)

— *We Need to Talk About Kelvin* (Faber and Faber, 2009)

Clarke, Arthur C., *The Other Side of the Sky* (Penguin, 1987)

Coase, Ronald, 'The Problem of Social Cost', *Journal of Law and Economics*, October 1960, pp. 1–44

Curzon, Paul, Peter McOwan and Jonathan Black, *Computer Science for Fun*, http://www.cs4fn.org/

Darwin, Charles, *The Origin of the Species* (Oxford World Classics, 2008)

Dawkins, Richard, 'The Ultraviolet Garden', Royal Institution Christmas Lecture No. 4, 1991

— *The Ancestor's Tale: A Pilgrimage to the Dawn of Evolution* (Phoenix, 2005)

— *The Blind Watchmaker* (Penguin, 2006)

— *The Greatest Show on Earth: The Evidence for Evolution* (Free Press, 2009).

Dennett, Daniel, *Consciousness Explained* (Back Bay Books, 1992)

Diamond, Jared, *Guns, Germs and Steel* (Vintage, 2005)

Eddington, Arthur, *The Nature of the Physical World* (1915)

Elbaz, David, et al., 'Quasar Induced Galaxy Formation: A New Paradigm?', *Astronomy and Astrophysics*, vol. 507 no. 3 (1 December 2009), pp. 1359–74; http://arxiv.org/abs/0907.2923

Elsasser, Walter, *Memoirs of a Physicist in the Atomic Age* (Watson, 1978)

Evans, Dylan, and Howard Selina, *Evolution: A Graphic Guide* (Icon Books, 2010)

Feresten, Spike, 'The Reverse Peephole', *Seinfeld* season 9 episode 12, 15 January 1998

Feynman, Richard, *QED: The Strange Theory of Light and Matter* (Penguin, 1990)

— *The Feynman Lectures on Physics*, edited by Robert Leighton and Matthew Sands, vols I and II (Addison-Wesley, 1989)

Fischbach, Gerald D., 'Mind and Brain', *Scientific American*, vol. 267 no. 3 (September 1992), p. 49

Ford, Kenneth, and John Wheeler, *Geons, Black Holes and Quantum Foam* (W. W. Norton, 2000)

Freud, Sigmund, *Civilisation and Its Discontents* (1929).

Germonpré, Mietje, et al., 'Fossil Dogs and Wolves from Palaeolithic Sites in Belgium, the Ukraine and Russia: Osteometry, Ancient DNA and Stable Isotopes', *Journal of Archaeological Science*, vol. 36 (2009), pp. 473–90

Gill-Mann, Murray, 'What Is Complexity?', *Complexity*, vol. 1 no. 1, 1995

Gleick, James, *Chaos* (Vintage, 1997)

Gould, Stephen Jay, *Wonderful Life* (Vintage, 2000)

Grabianowski, Ed, 'Why slime molds can solve mazes better than robots, www.io9.com, 12 October 2012 http://tinyurl.com/9ud95jx

Gray, John, *False Dawn: The Delusions of Global Capitalism* (Granta Books, 2009)

'Great Names in Computer Science' (http://www.madore.org/~david/computers/greatnames.html)

Gwynne, Peter, Sharon Begley and Mary Hager, 'The Secrets of the Human Cell', *Newsweek*, 20 August 1979, p. 48

Harford, Tim, *The Undercover Economist* (Random House, 2007)

Hartle, James, 'The Physics of Now', *American Journal of Physics*, vol. 73, issue 2 (February 2005), p. 101; http://arxiv.org/abs/gr-qc/0403001

Hartley, L. P., *The Go-Between* (Penguin Classics, 2004)

Heisenberg, Werner, *Quantum Theory* (1930)

— *Physics and Philosophy* (1963)

Hodges, Andrew, *Alan Turing: The Enigma* (Vintage, 1992)

How evolution REALLY works, http://tinyurl.com/brbpqod

Hutton, James, 'Theory of the Earth; or an Investigation of the Laws Observable in the Composition, Dissolution, and Restoration of Land upon the Globe', *Transactions of the Royal Society of Edinburgh*, vol. 1 (1788), pp. 209–304.

Johnson, George, *In the Palaces of Memory: How We Build the Worlds Inside Our Heads* (Vintage, 1992)

Jones, Terry, 'A Very Silly Way to Boil Water', *New Scientist*, 18 June 1987, p. 63

Kahlbaum, Georg W. A., and Francis V. Darbishire (eds), *The Letters of Faraday and Schoenbein, 1836–1862* (Williams and Norgate, 1899)

Kelly, Heather, 'OMG, the text message turns 20', CNN, 3 December 2012, http://tinyurl.com/cgoakdg.

Koene, Randal A., 'How to Copy a Brain', *New Scientist*, 27 October 2012, p. 26

Lane, Nick, *Life Ascending* (Profile, 2010)

Larkin, Philip, *High Windows* (Faber and Faber, 1967)

'Laws of Physics for Cats', http://www.funny2.com/catlaws.htm

Le Page, Michael, 'A Brief History of the Genome', *New Scientist*, 15 September 2012, p. 30

Leakey, Richard, and Roger Lewin, *Origins* (Penguin, 1982)

Leitch, Ilea, et al., 'Evolution of DNA Amounts Across Land Plants (Embryophyta)', *Annals of Botany*, vol. 95 issue 1 (January 2005), pp. 207–17

Lessing, Doris, *The Four-Gated City* (Flamingo Modern Classics, 2012)

Lewis, Gilbert Newton, *The Anatomy of Science* (Oxford University Press, 1926)

Lloyd, Seth, 'Ultimate physical limits to computation', *Nature*, vol. 406 no. 6799 (31 August 2000), p. 1047

Lockwood, Michael, *The Labyrinth of Time* (Oxford University Press, 2005)

Lorenz, Edward, *The Essence of Chaos* (Routledge, 1998)

Luther, Martin, *The Table Talk of Martin Luther*, translated by William Hazlitt (1872)

Maxwell, James Clerk, 'On the Motions and Collisions of Perfectly Elastic Spheres', *Philosophical Magazine*, January and July 1860

Médaille, John, 'Friends and Strangers: A Meditation on Money', *Front Porch Republic*, 20 January 2012; http://tinyurl.com/6q3pbsy

Minsky, Marvin, *The Society of Mind* (Simon & Schuster, 1988)

Misner, Charles, Kip Thorne and John Wheeler, *Gravitation* (W. H. Freeman, 1973)

Moore, Gordon, 'Cramming More Components onto Integrated Circuits', *Electronics*, vol. 38 no. 8 (19 April 1965)

Morran, Levi, et al., 'Running with the Red Queen: Host-Parasite Coevolution Selects for Biparental Sex', *Science*, 8 July 2011, vol. 333, pp. 216–18

Moulton, Forest Ray (ed.), *The Cell and the Protoplasm*, Publication of the American Association for the Advancement of Science, No. 14 (Science Press, 1940)

Norvig, Peter, 'Brainy Machines', *New Scientist*, 3 November 2012, p. vii

'The Origin of the Brain', http://tinyurl.com/d7sbhpk

'The Origin of Sexual Reproduction', http://tinyurl.com/ca8sjwg

Pamuk, Orhan, *The Museum of Innocence* (Faber and Faber, 2008)

Pinker, Steven, *The Better Angels of Our Nature: A History of Violence and Humanity* (Penguin, 2010)

Polanyi, Karl, *The Great Transformation* (Rinehart, 1944)

Pugh, George E., *The Biological Origin of Human Values* (Basic, 1977)

Rees, Martin, and Mitchell Begelman, *Gravity's Fatal Attraction* (Cambridge University Press, 2009)

Resnikoff, Howard, *The Illusion of Reality* (Springer-Verlag, 1989)

Ridley, Matt, *The Red Queen: Sex and the Evolution of Human Nature* (Penguin, 1994)

—— *The Rational Optimist* (Fourth Estate, 2010)

Ryan, Robert T., *The Atmosphere* (Prentice Hall, 1982)

Rutherford, Adam, *Creation* (Penguin, 2013)

Sagan, Carl, 'Wonder and Skepticism', *Skeptical Enquirer*, vol. 19 no. 1 (January–February 1995)

Scharf, Caleb, *Gravity's Engines* (Allen Lane/Penguin, 2012)

Schulman, L. S., 'Source of the Observed Thermodynamic Arrow', *Journal of Physics: Conference Series*, vol. 174 no. 1 (2009), 12,022

Shipman, Pat, 'Man's Best Friends: How Animals Made Us Human', *New Scientist*, 31 May 2011, p. 32

Sokal, Alan, 'A Physicist Experiments with Cultural Studies', *Lingua*

Franca, May/June 1996; http://tinyurl.com/mvow

Stenger, Victor, *Timeless Reality: Symmetry, Simplicity, and Multiple Universes* (Prometheus Books, 2000)

Stiglitz, Joseph, 'Inequality is holding back the recovery', in the series 'The Great Divide', *New York Times*, 19 January 2013, http://tinyurl.com/ aqxj9ro

Stoppard, Tom, *Hapgood* (Faber and Faber, 1988)

Stringer, Chris, *The Origin of Our Species* (Allen Lane, 2011)

— and Peter Andrews, *Complete World of Human Evolution* (Thames & Hudson, 2011)

Taleb, Nassim, *The Black Swan: The Impact of the Highly Improbable* (Penguin, 2008)

Tegmark, Max, 'Is the "Theory of Everything" Merely the Ultimate Ensemble Theory?', *Annals of Physics*, vol. 270, issue 1 (20 November 1998), pp. 1–51

Thomas, Lewis, *The Lives of a Cell* (Penguin, 1978)

— *The Medusa and the Snail* (Penguin, 1995)

Tibballs, Geoff, *The Mammoth Book of Zingers, Quips, and One-Liners* (Running Press, 2004)

Twain, Mark, *Following the Equator* (1897)

Tyndall, John, *In Forms of Water in Clouds and Rivers, Ice and Glaciers* (1872)

Tyson, Neil deGrasse, *Death by Black Hole: And Other Cosmic Quandaries* (W. W. Norton, 2007)

Van Valen, Leigh, 'A New Evolutionary Law', *Evolutionary Theory*, vol. 1 (1973–76), pp. 1–30

Watson, James, *Discovering the Brain* (National Academy Press, 1992)

— *The Double Helix* (Phoenix, 2010)

Wilde, Oscar, *The Picture of Dorian Gray* (1890)

— *De Profundis* (1905)

Wilson, Edward O., *Consilience* (Vintage, 1999)

— *The Social Conquest of Earth* (W. W. Norton & Co., 2012)

Wolpert, Lewis, 'Shaping Life', *New Scientist*, 1 September 2012

Young, Louise, *Earth's Aura* (Random House, 1977)

FURTHER READING

PART ONE: How We Work

David Barash, *Homo Mysterious: Evolutionary Puzzles of Human Nature*
Charles Darwin, *The Origin of the Species*
Richard Dawkins, *The Blind Watchmaker*
Dylan Evans and Howard Selina, *Evolution: A Graphic Guide*
How evolution REALLY works, http://tinyurl.com/brbpqod
Nick Lane, *Life Ascending*
Chris Stringer, *The Origin of Our Species*
Chris Stringer and Peter Andrews, *Complete World of Human Evolution*
James Watson, *The Double Helix*

PART TWO: Putting Matter to Work

Mark Buchanan, *Forecast: What Physics, Meteorology and the Natural Sciences Can Teach Us about Economics*
Ha-Joon Chang, *23 Things They Don't Tell You About Capitalism*
Jared Diamond, *Guns, Germs and Steel*
Richard Feynman, *QED: The Strange Theory of Light and Matter*
Richard Feynman, Robert Leighton and Matthew Sands, *The Feynman Lectures on Physics*, vol. II
'Great Names in Computer Science', http://www.madore.org/~david/computers/greatnames.html
Tim Harford, *The Undercover Economist*
Andrew Hodges, *Alan Turing: The Enigma*
Steven Pinker, *The Better Angels of Our Nature: A History of Violence and Humanity*
Matt Ridley, *The Rational Optimist*

PART THREE: Earth Works

James Gleick, *Chaos*
Edward Lorenz, *The Essence of Chaos*
Louise Young, *Earth's Aura*

PART FOUR: Deep Workings

Peter Atkins, *Four Laws that Drive the Universe*
Sander Bais, *Very Special Relativity*
Marcus Chown, *Quantum Theory Cannot Hurt You*
— *We Need to Talk About Kelvin*
Richard Feynman, *QED: The Strange Theory of Light and Matter*
Michael Lockwood, *The Labyrinth of Time*
Charles Misner, Kip Thorne and John Wheeler, *Gravitation*
Victor Stenger, *Timeless Reality: Symmetry, Simplicity, and Multiple Universes*

PART FIVE: The Cosmic Connection

Marcus Chown, *The Never-Ending Days of Being Dead*
Martin Rees and Mitchell Begelman, *Gravity's Fatal Attraction*
Caleb Scharf, *Gravity's Engines*

INDEX

absolute zero, 330, 392 n. 7
AC (alternating current), 131, 133, 368–9 n. 19
acceleration, 263–70, 303, 347, 384 n. 2, 384 n. 6 *infra*; *see also* gravity
acetycholine, 75
acidity, 29
acids, 6, 147; fatty acids, *see* lipids; *see also* amino acids
Acorn Computers, 147
Ada computer language, 370 n. 3
Adams, Douglas, 301, 323
Adder's-tongue fern, *see Ophioglossum vulgatum*
adenine, 42
aeroplanes, 115, 141, 374–5 n. 9; Boeing 747, 197
Afar, Ethiopia, 187
Africa, 96–7, 100, 113, 114, 183, 185–7, 268, 302; *see also* South Africa
Agassiz, Louis, 191
agriculture, *see* farming
air, 23, 27, 28–30, 252; convection, 197–9, 206, 374 n. 8, 375 nn. 10, 12; 376 n. 16; pressure, 215–16; pressure equalisation, 199; resistance, 263; in a thunderstorm, 125–8, 367 nn. 7, 9; *see also* Earth: atmosphere, Hadley circulation cells, vortices, weather systems, wind
alchemy, 283
alcohol, 25
algae, 356 n. 6; blue-green, 195, 205
alleles, 59
Allen, Tim, 137
Allen, Woody, 267
alpacas, 117
Alpha Centauri, 301, 302

alpha particles, 284, 286
altruism, 65
alveoli, 28–9
Amazon.com, 127
Americans, Native, 118; *see also* Aztecs, Incas
Americas, 114–17; *see also* North America, South America
amino acids, 6, 42, 75
amygdala, *see* limbic system
ancestry, 361–2 n. 13
Anderson, Philip, 311
Andes, 117, 188
Andromeda Galaxy, 302, 326
angular momentum, 238; law of conservation of, 238, 243
animals, 296; diseases, 118; domestication, 37, 101, 112–13, 116–17, 162; extinction, 117, 119; and food, 25; interbreeding, 44; mitochondria, 10; multicellular, 14, 74; as pets, 113; as prey, 113; reproduction, 38, 63; respiration, 23, 25, 28, 196; sight, 47; single-celled, 16; stamina, 93; variation, 37, 43, 90; *see also* bones, clothing, meat, natural selection, nervous system
Antarctic ice sheet, 207
Antarctica, 96, 204, 205
antibiotics, 16, 45, 115; resistance to, 45
anticyclones, *see* weather systems
antimatter, 290
antiparticles, 290–91, 350; electrical charge, 290; spin, 290
anti-quarks, 291
ants, 156
apes, 89, 92, 111, 113, 156, 297, 315
Apollo 15, 263
apoptosis, 27

archaea bacteria, 9–10, 356 nn. 5, 6
Arctic Sea, 205
argon, 195
ARM, 147
Armstrong, Neil, 119
arsenic, 146
art and artists, 109, 113
artificial selection, 108
asexual reproduction, 52–4, 56, 60
Asia, 96, 107, 114–16, 118
Aspect, Alain, 239
asteroids, 195
astronomy and astronomers, 10, 14, 39, 268, 271, 274, 275, 288, 301–3, 316, 325, 345, 347–9
@Katharine_T29m, 381 n. 7
@MrDFJBaileyEsq, 382 n. 17
Atkins, Peter, 213, 215
Atlantic Ocean, 96, 97, 185–6, 188, 201, 204; see also mid-Atlantic Ridge, transatlantic cables
atomic clock, 266, 273
atoms, 23, 26–8, 31, 73, 124–5, 128, 129, 133, 145–6, 216–18, 279–98, 343, 350, 365–6 n. 2, 368 n. 15, 375–6 n. 13, 378 n. 4, 381 n. 5, 388 n. 2, 389 n. 10, 392 n. 7, 393 n. 12, 394 n. 14; alignment, 186; doping, 146, 147; electrically charged, 135, 137, 363 n. 6; nucleus, 23–4, 26, 124, 129, 240–42, 244, 252, 283–5, 286, 288, 296, 328, 358 n. 6, 366 n. 3, 386 n. 14, 388 n. 2, 389 n. 10, 392 nn. 6, 8; 394 n. 14; and quantum theory, 227–9, 231, 233, 235–7, 240–44, 381 n. 5; structure, 23–4, 283–5, 357 n. 3; see also electrons, leptons, molecules, muons, neutrons, neutrinos, photons, protons, quarks
ATP (adenosine triphosphate), 10, 27, 137
Augarten, Stan, 145
austerity, 177
Australia, 114–18, 196; fossils, 8
australopithecines, 91, 94, 120
Australopithecus afarensis, 91
Australopithecus anamensis, 90
axes, 94–5

axons, see nerve cells
Aztecs, 117

Babbage, Charles, 370 n. 3
Bacon, Francis, 185
bacteria, 11, 16–17, 18, 25, 35, 45, 52, 56, 60, 72, 74, 75, 356 nn. 5, 6; 360 n. 2, 361 n. 6; reproduction, 43, 45, 355 n. 3; structure, 6; see also archaea bacteria
balance, 79, 81
balloon, electrically charged, 127–8, 367 nn. 9, 10
balloons, hot-air, 196
Bangladesh, 169
banking, 151, 159–61; assets, 161; borrowing, 160, 163; defaulting, 160; interest, 160; investment, 170; lending, 160–61; merchant, 164
banking crisis (2008), 160, 161, 170, 175, 176, 371 n. 2
Bardeen, John, 370 n. 9
Barings Bank, 164
Barry, Dave, 121, 309
baryons, 291
battery, 27–8, 366–7 n. 5
BBC, 329, 351
Beagle, HMS, 35
bees, 156
Begley, Sharon, 78
Bekenstein, Jacob, 350
bell curve, see Gaussian distribution
Bell Laboratories, 370 n. 9
Bern, 265
Berners-Lee, Tim, 80
Besso, Michele, 306
Bezos, Jeff, 127
big bang, 222, 274, 288, 290, 291, 293–4, 303, 306, 308, 315, 319, 326–37, 380 n. 13, 381 n. 5, 388 n. 2, 389 n. 10, 391 n. 4, 391–2 n. 5, 394 nn. 14, 16, 18; 394–5 n. 19; inflation, 330, 335, 336
binary, 142–4, 145
binary pulsar, the, see PSR B1913+16 (binary pulsar)
Binnig, Gerd, 386 n. 3
biofilms, 8

biology and biologists, 5, 6, 10, 13–15, 18, 25, 26, 29, 36, 38, 40–42, 45, 46, 51, 52, 54–6, 58, 60, 65, 77, 85, 101, 107, 118, 137, 155, 189, 215, 296, 308, 336, 355–6 n. 5, 356 n. 11, 359 n. 7, 361 n. 6
bipedalism, 91–4
bisexuality, 65
bison, 117
Black, Jonathan, 370 n. 6
black holes, 18, 266, 270, 275, 277, 341–52, 384 n. 4, 395 n. 3, 395–6 n. 5; entropy, 351; singularity, 344, 350; stellar-mass, 345–6; supermassive, 345–9; see also Cygnus X-1, Sagittarius A*
Blair, Tony, 371 n. 2
Blake, William, 279, 311
blood, 5, 358 n. 10
blood cells, 14, 15, 28–9, 43, 64–5
blood clot, 80
blood group, see genes
blood vessels, 28–9
Blumenthal, Sidney, 172
body temperature, 79
Bohr, Niels, 225, 245, 279
Bollier, David, 176, 177; Silent Theft, 176
Boltzmann, Ludwig, 307
Boltzmann's constant, 389 n. 9
Bondi, Hermann, 259, 391–2 n. 5
bones, 91, 94, 96
Boolean logic, 390 n. 5
boom and bust, 163–4
boron, 146
bosons, 245, 289, 291–2, 387–8 n. 15; spin, 387–8 n. 15; vector, 289; see also Higgs boson
Bostrom, Nick, 369 n. 2
bottom muscle, see gluteus maximus
Boyle's law, 228
Bragg, William, 230
brain, human, 60, 69–86, 90, 94, 296, 297, 303; brainstem, 78; cerebellum, 78; copying, 363 n. 13; emotion, 79; energy use, 77–8, 93, 99; grey matter, 363–4 n. 17; memory, 79; reptilian, 79; size, 101; weight, 78; white matter, 363–4 n. 17; see also limbic system, neocortex

brain cells, see neurons
brainstem, see brain
Brattain, Walter, 370 n. 9
breathing, 28–9, 79
breeding, deliberate, 37
Britain, 188
Brno, 40
Broome, Western Australia, 391 n. 9
Brown, Robert, 12, 356 n. 11
Brunel, Isambard Kingdom, 372 n. 4
Buchanan, Mark, 172, 174–5
Bushmen (Kalahari), 93
Butler, Samuel, 52
butterfly effect, see chaos theory
Byron, George Gordon, Lord, 370 n. 3

Caenorhabditis elegans, 56, 78, 363 n. 13
calcium, 73
calculus, 143
Caledonian mountains, 184
California, University of, at Santa Barbara, 308
California Institute of Technology, Pasadena, 377–8 n. 23
Cambridge, 147
camels, 118
Canada, 201, 205
cancer-causing chemicals, 43
capillaries, 65
capital, 167
capitalism, 165–78; free-market, 168; profit, 177–8; see also market, the
carbon, 28, 30, 348
carbon dioxide, 28–30, 190, 195, 205–8, 358 n. 11, 378 n. 1
Carlyle, Thomas, 297–8
carnivores, 25
Carnot, Sadi, 379 n. 7
Carroll, Lewis, 55; Alice in Wonderland, 55, 227; Through the Looking-Glass, 55
cars, 31, 115, 148, 317
Cassini spacecraft, 144
cattle, 118
cats, 57, 314; sabre-toothed, 96
cave-dwelling, 98
CDO (Collateralised Debt Obligation), 175

CDO-squared, 175

cells, 5–19, 27, 41–2, 63, 110, 296; chemical signalling, 8, 73, 75, 363 n. 8; complexity, 9; cooperation, 8–9, 12, 13; communication, 15, 72–5; creation, 57, 58, 62, 220; and dehydration, 8; division, 15, 359–60 n. 8; and DNA, 42, 56, 57, 61, 64; electrical charge, 73–6, 137; fusion, 60–62; heat-sensitive, 47; integration, 12; ion channels, 73–5; light-sensitive, 47; membrane, 7–8, 11, 12, 14, 27, 61, 73, 220, 356 n. 5, 358 n. 7; microtubules, 12–13; nucleus, 12–13, 14, 42, 74, 110, 356 n. 12; 361 n. 10; protection, 8; and proteins, 11–12, 358 n. 5; reconstitution, 363 n. 5; size, 9; structure, 7, 42, 358 n. 5; transportation system, 13, 14; visual appearance, 5; voltage-gated ion channels, 73–4; see also apoptosis, blood cells, cytoplasm, endoplastic reticulum, Golgi apparatus, gonad cells, Krebs cycle, lysosomes, mitochondria, neurons, organelles, prokaryotes, ribosomes, vesicles

Celsius scale, 379 n. 10

centrifugal force, 269, 372–3 n. 5

cereal crops, 107–9

cerebellum, see brain

cerebrovascular fluid, 79

cerebrum, see neocortex

chalk, 208

Chandrasekhar, Subrahmanyan, 341

Chang, Ja-Hoon, 169, 171, 172, 175, 176; 23 Things They Don't Tell You About Capitalism, 169

chaos theory, 201

chemistry and chemists, 24, 28, 136–7, 207, 215, 242–3, 283, 296

Cherenkov light, 387 n. 7

Chicago School, 173

chickens, 118

child bearing, 66

child labour, 168

child rearing, 65–7

Chile, 188, 189

chimpanzees, 36, 45, 58, 89, 92, 93; DNA, 45, 58, 89–90; see also hominins

China, 96, 112

chips, see microchips

chlorophyll, 30

chloroplasts, 356 n. 6

chromatin, 361 n. 10

chromosomes, 41–2, 56–9, 62, 361 n. 10; 362 n. 15; number, 57; X, 59; Y, 59; see also SRY

Church, George, 359–60 n. 8

Churchill, Winston, 33, 51, 360 n. 1

cilia, 74

civilisation, 105–20, 181

Clarke, Arthur C., 337; 2001: A Space Odyssey, 94

Clarkson University, New York, 388 n. 8

Clausius, Rudolf, 222

climate, 203–4

climate change, 92, 98, 108, 119

Clinton, Bill, 172, 371 n. 2

clothing, 98, 100

coal, 30, 114, 131, 163, 172, 206, 370 n. 5, 378 n. 1

Coase, Ronald, 173

cobbles, 94–5

colour, 273, 281, 297; spectrum, 274, 385 n. 11

comets, 195, 360 n. 2

commerce, see trade

commodities, 157–8, 177

communism, 165, 167, 178

compass, magnetic, 129, 368 n. 12

computers, 31, 83, 85, 115, 119, 139–49, 216, 227; adders, 145, 146; AND/OR gates, 145; circuits, 77; desktop, 147; gates, 144–5; halting problem, 142–3; integrated circuits, 146, 370 n. 10; laptops, 149; logic gates, 77, 145, 146; magnetic medium, 144; memory, 81; PCs, 147; power, 148–9, 369 n. 2; programming, 83, 89, 141–3; speed, 149; see also microchips, silicon, transistors, uncomputability

Conan Doyle, Arthur, 363 n. 1

conductors: electricity, 145, 146, 366–7 n. 5, 368 n. 15; heat, 374 n. 8; see also wire coil, conducting conic sections, 385 n. 10

Conklin, Edwin Grant, 296

consciousness, 79

continental crust, 188–90, 373 n. 7

continental drift, 185

continents, 185, 188, 204

cooking, 99

Copernicus, Nicolaus, 39

copper, 368 n. 15

Coriolis force, 375 n. 10

corpus calossum, see neocortex

cosmic background radiation, 222, 327, 330–31, 333, 334, 388 n. 2, 391 nn. 3, 4; 391–2 n. 5, 392 n. 8, 394 n. 14; temperature, 330, 334–5

cosmic rays, 386 n. 14

cosmic speed limit, 239, 255, 262, 303, 392 n. 9

cosmology and cosmologists, 288, 323–39, 381 n. 5; precision cosmology, 394 n. 18

craft and craftsmen, 110, 114, 115

creationism, 35–6, 38–9, 183, 313, 390 n. 4

credit cards, 162

credit-rating agencies, 161

Crick, Francis, 359 n. 6

Cringely, Robert X., 148, 371 n. 14

Cro-Magnons, 101

crops, see farming

crystal, 235, 236, 381–2 n. 8

cuneiform, 111

curiosity, 71, 86

Curzon, Paul, 370 n. 6

cyanobacteria, 119, 195, 205, 356 n. 6

cyclones, see weather systems

Cygnus X-1, 345

cystic fibrosis, 43

cytoplasm, 7, 8, 9, 13

cytosine, 42

Dallas, Texas, 126

Dalrymple, David, 363 n. 13

dark energy, 292, 333–4, 336, 380 n. 16, 393–4 n. 13

dark matter, 292, 335, 336, 394 n. 15

Darwin, Charles, 18, 36–41, 45–6, 84–5, 87, 90, 91, 176, 183, 189, 336, 358–9 n. 2, 359 nn. 3, 4; The Origin of Species, 39

Davisson, Clinton, 235, 381–2 n. 8

Dawkins, Richard, 11, 38, 46, 47, 51, 53, 64, 91; The Blind Watchmaker, 51

DC (direct current), 131, 133

de Broglie, Louis, 230, 382 n. 14

dead, burial of, 99

decay, 307, 332, 393 n. 12

decimal, 143

Deep Time, 184

democracy, 178

Democritus, 281–2, 285, 292, 295, 386 n. 3

dendrites, see neurons

Dennett, Daniel, 78

density, 282, 328, 335, 344, 380 n. 13

design, illusion of, 35, 46

deuterium, 373–4 n. 2

developing countries, 170–71

Diamond, Jared, 115, 116, 118

diarrhoea, 16

Dickens, Charles, 111

Dickinson, Emily, 71

Diller, Phyllis, 49

dinosaurs, 36, 125, 326

diploid, 61

Dirac, Paul, 315

DNA (deoxyribonucleic acid), 7–8, 10–12, 14, 17, 18, 19, 42, 51–2, 54, 56, 58, 61–2, 64, 78, 85, 89, 189, 296, 356 nn. 5, 12; 361–2 n. 13; copying, 15, 43, 53, 359 n. 6; crossover, 62; information storage, 355 n. 4; 359–60 n. 8; human, 42–3, 45; mitochondrial, 361–2 n. 13; mutation, 43, 53; packaging, 361 n. 10; plant, 108; protein expression, 9, 12–13, 42–3; redundancy, 43; structure, 10, 42, 56, 355 n. 4; supercoiling, 361 n. 10; see also adenine, chromosomes, cytosine, guanine, human beings, thymine

Dobzhansky, Theodosius, 38

dogs, 57, 113, 365 n. 6

domestic appliances, 123, 130, 132

dopamine, 75

doping, see atoms: doping

doppelgänger, 338–9
dot-com bubble, 164, 371 n. 2
Drake Passage, 204
Durant, Will, 181

Eagle pub, Cambridge, 359 n. 6
Earth, 183–4; age, 184; atmosphere, 25, 97,
    190, 193–209, 252–3, 375–6 n. 13, 378
    n. 1; biosphere, 221; core, 190–91; crust,
    174, 184–90; equator, 98, 197, 198,
    201–2, 204, 205, 374 nn. 5, 8; 375 nn. 10,
    11; equatorial regions, 196, 367 n. 7, 374
    n. 5; formation, 372–3 n. 5; gravity,
    269–70, 304, 314; habitability, 190, 195;
    heat, 189–91, 196–8, 201–2, 204–6, 208,
    215, 219, 374 nn. 5, 8; 376 n. 16, 378 n. 1;
    interior, 189; magnetic field, 129, 186;
    magnetic poles, 186; mantle, 187–91;
    northern/southern hemiheres, 197,
    199–200, 202, 203, 301, 375 n. 11, 376
    n. 16; orbit, 97, 202, 203, 249, 262, 306,
    325, 364 n. 8, 377–8 n. 23; polar band,
    198, 375 n. 12; polar regions, 198, 374
    n. 5; poles, 196–8, 201–2, 204, 205, 374
    n. 8, 375 nn. 10, 11, 12; 376 n. 17;
    rotation, 196–200, 374–5 n. 9, 375
    nn. 10, 12; seasons, 202, 376 nn. 16, 17;
    size, 195; solstices, 202; spin axis, 97,
    202, 203, 364 n. 8, 376 n. 17; strata, 185;
    stratosphere, 204; subsolar point, 202;
    temperature, 195, 196, 200, 202, 206–7,
    220, 376 n. 16, 379 n. 10; tropical band,
    199, 375 n. 12; troposphere, 200, 204,
    206; see also air, climate, continental
    crust, continental drift, Milanković
    cycles, oceanic crust, Snowball Earths,
    tectonic plates, weather systems
earthquakes, 174, 189, 190
ecology, 177
economics, 157–8, 164, 172–5, 177, 371 n. 2;
    see also econophysics, inflation, protec-
    tionism, stagnation, subsidy, trade
econophysics, 175
eddies, see vortices
Eddington, Arthur, 219, 259, 268
Edison, Thomas, 127, 132, 367 n. 8

efficiency, see energy
egg(s), see ovum/ova
Einstein, Albert, 129, 134, 149, 225, 231–2,
    239, 247, 249–51, 254, 256–8, 261–5,
    267–9, 271–7, 290, 302–6, 309, 317–18,
    334, 343, 344, 356 n. 11, 368 n. 12, 369
    n. 20, 381 n. 5, 384 n. 6 infra, 385 n. 8,
    392 n. 9, 393 n. 11, 393–4 n. 13
electric current, 126, 129–32, 144, 216,
    366–7 n. 5; voltage, 132–3, 144, 145, 368
    n. 18; see also AC, DC
electric field, 27, 128–31, 133, 134–5, 137, 368
    nn. 15, 17; 369 n. 20, 383–4 n. 5
electric force, 123–7, 136–7, 291, 365–6
    n. 2, 368 n. 11
electric generator, 131, 133
electric motor, 130, 133
electricity, 121–37, 145, 366–7 n. 5, 382
    n. 10; charge imbalance, 125–7, 367
    nn. 7, 9; 368 n. 13; domestic supply, 121,
    131–2, 136, 368 n. 17; insulation, 146,
    363–4 n. 17, 368 n. 17; long-distance
    transmission, 131–3; within cells,
    73–6, 363–4 n. 17; see also conductors,
    domestic appliances, National Grid,
    power stations, semiconductors, wire
    coil
electrodynamics, 135
electromagnetic fields, 134
electromagnetic force, 289, 291, 343
electromagnetic spectrum, 385 n. 11
electromagnetic waves, 135, 240
electromagnetism, 136, 240, 378 n. 4;
    Maxwell's theory of, 240
electron degeneracy pressure, 382 n. 15
electron-neutrinos, see neutrinos
electron waves, 230, 235, 241–3, 366 n. 3,
    382 n. 15, 383 n. 19, 387 n. 11
electrons, 21, 23–4, 26–8, 30, 31, 124–6,
    128, 131–3, 144–6, 148, 216, 283–8,
    292–4, 328, 357 nn. 1, 3, 4; 358 nn. 6, 7;
    365–6 n. 2, 366 nn. 3, 4; 366–7 n. 5, 367
    n. 9, 368 nn. 15, 17; 383–4 n. 5, 389 n. 10,
    392 n. 8, 394 n. 14; combination, 24, 26,
    388 n. 2; indistinguishability, 243, 383
    n. 19, 387 n. 11; orbit, 241–2, 283, 285,

287; positively/negatively charged, 123–5, 129, 136–7, 146, 283–4, 366–7 n. 5, 367 n. 10, 368 n. 13; and quantum theory, 231, 234–6, 238–40, 243–4, 381–2 n. 8, 382 n. 15; shell arrangement, 28, 244, 357 n. 3; spin, 238–9, 243–5, 287, 382 nn. 9, 10, 383 n. 19, 387 n. 11

electroweak force, 291

elements, 283

Eliot, T. S., 222

ellipse, 385 n. 10

embryo development, 15, 42, 59–60, 63, 89–90, 359 n. 7

emergence, 295, 297

employment, 164

endoplastic reticulum, 11

energy, 273, 288, 329, 337–8, 350, 385 n. 8; chemical, 217, 255, 317; density, 334, 393 n. 11, 393–4 n. 13; dissipation, 368 n. 17; efficiency, 53, 78, 379 n. 7; electrical, 24, 125–6, 149, 217, 255; explosive, 124–5, 366 n. 4; generation, 6, 7, 10–11, 14, 21; geochemical, 358 n. 11; gravitational, 24, 242, 272, 273; heat, 26, 127, 130, 133, 216–21, 255, 257–8, 262, 272, 368–9 n. 19, 378 n. 3, 379 n. 7; law of conservation of, 217, 255, 286, 317, 318, 366 n. 4, 379 n. 6, 393 n. 10; liberation of, 24–6, 29, 31; light, 24, 30, 127, 130, 149, 217, 241, 242, 272, 379 n. 11; loss, 240; lowest, 357 n. 3; magnetic, 377 n. 19; and mass, 149, 256–7, 262, 267, 286, 290, 292–3, 333, 344, 350, 384 n. 5; minimisation, 24, 242; of motion, 217–18, 255–7, 290, 317, 357 n. 1, 380 n. 14, 392 n. 6; nuclear, 345; orbital, 276; per person, 162–3; requirement, 9; solar, 196, 221, 358 n. 11, 378 n. 1; sound, 220, 256, 272; source, chemical, 23; surplus, 24–6, 241–2, 392 n. 6; transformation, 24; ultra-high, 293; and work, 24, 25, 357 n. 1; see also dark energy, sunlight

'Enigma' code, 142

entropy, 218–21, 351, 379 n. 8, 380 nn. 12, 14; 389 n. 9

epigenetics, 64

'epoch of last scattering', see 'last scattering, epoch of'

ESA (European Space Agency), 197

eukaryotes, 10–13, 17, 60, 356 n. 5; 361 n. 10; cytoskeleton, 12

eukaryotic cells, 10, 61, 356 n. 6; fusion, 61

Euphrates river, 107

Europe, 96, 101, 114–18, 201, 204, 205

evaporation, 193

event horizon, 277, 343–4, 350–52

Everest, Mount, 386 n. 12

evolution, 33–47, 62, 63, 85, 108, 155, 176, 189; see also artificial selection, natural selection

evolution, human, 87–101

exchange particle, 289

exchange rate, 154, 155

extinction, 56, 64, 98, 117, 119, 125, 174, 309

Everything, Theory of, 294–5, 298

eye, 46–7, 228, 303, 327–8; retina, 237

eye colour, see genes

factories, 163, 164

Falklands, 186

far infrared, 206, 222, 379 n. 11, 391 nn. 3, 4

Faraday, Michael, 130, 131, 369 n. 21

Faraday rotation, 369 n. 21

farming, 107–9, 113, 116–18, 120, 161, 365 n. 2 supra

fascism, 167

fat, 26, 358 n. 9

Federal Reserve Board (US), 171

Ferguson, Niall, 151

fermentation, 25

fermions, 245, 291–2, 387 n. 11, 387–8 n. 15; spin, 387–8 n. 15

Fertile Crescent, 107, 108, 112, 116, 117

fertilisation, internal, 63

feudalism, 167

Feynman, Richard, 135, 136, 240, 244, 281, 313; The Feynman Lectures on Physics, 279

finches, 36–7, 46

fire, 98–9

Fischbach, Gerald D., 80

Fischer, Irving, 173
'Fish' code, 142
Flores, Indonesia, 96
flu, 118
flying foxes, 35
fMRI (functional magnetic resonance imaging), 84
food, 10, 16, 21, 23, 26, 28–30, 38, 53, 91, 137, 358 n. 9; chain, 25; and farming, 109, 113; production, 113, 114, 116, 118; sponges, 72; surplus, 109–10
force-carrying particles, 289–91
Ford, Henry, 151
foreign exchange, 174
forgetting, see neurons
fossil evidence, 6, 8, 35–6, 47, 90–91, 94, 96, 97, 101, 183, 185, 365 n. 6; see also stromatolites
fossil fuels, 30–31, 163, 172, 206; see also coal, oil
Franklin, Benjamin, 111, 366–7 n. 5
Freud, Sigmund, 105, 111
friction, 220, 231
Friedman, Milton, 173
frogs, 35
fruit flies, 41
fuel, 23, 58, 357 n. 2
fuel cells, 31
fungi, 14, 16, 35, 52, 356–7 n. 15
fur, 92–3
futures market, 174

GABA (gamma-aminobutyric acid), 75
Galápagos archipelago, 36–7, 44, 46
galaxies, 289, 315, 326, 328–9, 334–8, 345–8, 389 n. 10, 391 n. 1, 391–2 n. 5, 394 nn. 14, 15; ancient, 302; clusters, 315; distant, 274–5, 337; expansion, 274, 329; formation, 334–5, 337–8, 348–9, 372–3 n. 5, 391 n. 1, 391–2 n. 5, 394 n. 14; and gravity, 289, 292, 314–15, 326, 333, 389 n. 10; as image, 5, 15–16; intelligent life, 14; light from, 274–5, 328, 337, 345; number, 78, 326; radio emission, 347; recession, 326, 329; spiral, 394 n. 15; see also Andromeda

Galaxy, black holes, dark matter, jets, Milky Way Galaxy, quasars, stars
Galbraith, John Kenneth, 164, 165
Galileo Galilei, 250–51, 263, 374–5 n. 9
gallium, 146
games consoles, 147
gametes, 57–8, 63; fusion, 58, 362 n. 15
gamma rays, 385 n. 11, 392 n. 6
gases, 200, 282, 296–7; inter-stellar, 346–8; pressure, 375–6 n. 13
gathering, 107, 110; see also hunter-gatherers
Gauguin, Paul, 118
Gaussian distribution, 173
GDP (Gross Domestic Product), 177
Geiger, Hans, 284, 286
Gell-Mann, Murray, 295
genes, 17–19, 41, 42, 46, 52, 54, 58–62, 64–5, 66, 89–90, 296, 362 n. 15; activation, 359 n. 7; blood group, 89; combination, 44, 56; copying, 61; dominant, 59; expression, 89; eye colour, 89; human, 17; microbial, 17; mutation, 18, 43–6, 54, 58; recessive, 59; regulatory, 14, 90; sequence, 361 n. 6; see also alleles, diploid, haploid
genetic code, 42
genetic defects, 66
genetic engineering, 108, 112, 120
genetics, 41, 85, 189; see also DNA
Geneva, 256, 289, 347
genitalia, 63; insect, 44
genius, 361 n. 6
genome, human, 17
geology, 181–91
geometry, 390; see also space–time: warpage
Georgia, 96
Germany, 158
Germer, Lester, 235, 381–2 n. 8
Germonpré, Mietje, 365 n. 6
gibbons, 92
Gladstone, William, 130
glial cells, 363–4 n. 17
Global Positioning Satellites, 304; orbit, 304

global warming, 171–2, 204, 206–8, 378 n. 1
glucose, 26
gluon field, 256, 290
gluons, 289
glutamate, 60, 72, 75
gluteus maximus (bottom muscle), 91, 93
goats, 112
Gödel, Kurt, 370 n. 4
gold, 147, 282, 284; as currency, 159, 160
Gold, Thomas, 391–2 n. 5
Golgi apparatus, 11
gonad cells, 60, 63
gorillas, 35
Gould, Stephen Jay, 12
government, 110
Grahame, Kenneth, 363 n. 2
granite, 188, 373 n. 7
gravitational force, 262, 289, 293
graviton, 289, 293
gravity, 123–5, 137, 203, 216, 261–75, 277, 290, 292–3, 303, 304, 306, 308, 313–14, 333–5, 343, 344, 349, 365 n. 1 *infra*, 365–6 n. 2, 372–3 n. 5, 377 n. 21, 380 n. 13, 382 n. 15, 382–3 n. 18, 384 n. 4, 384 n. 6 *infra*, 385 nn. 8, 9, 10; 387 n. 12, 389 n. 10, 391 n. 8, 393 n. 11, 393–4 n. 13, 394 nn. 14, 15; 395–6 n. 5; Newton's theory of, 231, 261, 262; and planets, 23, 365–6 n. 2, 376 n. 17, 377–8 n. 23; repulsive, 331, 333–4, 380 n. 16, 393 n. 11; universal law of, 262, 270, 271, 306, 313–15, 390 n. 3; *see also* acceleration, galaxies: and gravity, red shift: gravitational, waves: gravitational
Great Eastern, 372 n. 4
Greece, 159
Greeks, 114–15, 271, 281, 313
Greene, Graham, 299
greenhouse gases, 190, 205–8, 377 n. 22, 378 n. 1
Greenland, 189, 205; ice sheet, 207
Greenspan, Alan, 171, 175
guanine, 42
guns, *see* weapons
Guth, Alan, 331
Gwynne, Peter, 14

habitat destruction, 171
Hadley, George, 196
Hadley circulation cells, 196, 197, 201, 375 n. 12
haemoglobin, 29
hair, 93
Haines, John, 327
half-life, 383 n. 2
hallucinations, 76
Hamilton, Andy, 385 n. 9
Hannibal, 301
haploid, 61
Hartle, Jim, 308–9
Hartley, L. P., 184, 306
Hartley 2, Comet, 373–4 n. 3
Harvard Medical School, 359–60 n. 8
Harvard University, 363 n. 13
harvesting, 108
Hawaii, 191
Hawking, Stephen, 87, 294, 350, 395–6 n. 5
Hawking radiation, 350–51, 395–6 n. 5
Hawthorne, Nathaniel, 121
H-bomb, *see* nuclear bomb
heart, 29, 358 n. 10
heart attack, 29
heart rate, 79
heat, *see* energy: heat
Heat Death, 221–2
heat loss, 92
heavy water *see* deuterium
Heidelberg, Germany, 97
Heisenberg, Werner, 229
Heisenberg Uncertainty Principle, 382 n. 15
helium, 285, 288, 328, 377 n. 21, 392 n. 6, 394 n. 16
heroin, 168
Hertz, Heinrich, 135
heterosexuality, 65
Higgs, Peter, 383–4 n. 5
Higgs boson, 289–91
Higgs field, 289, 383–4 n. 5
Higgs mechanism, 256
Higgs particle, 256
Himalayas, 97, 184, 188
hippocampus, *see* limbic system
histones, 361 n. 10

holograms, 343, 349–52
hominins, 90, 92, 94, 96–101, 364 n. 2; *see also* australopithecines, *Australopithecus afarensis*, *Australopithecus anamensis*, chimpanzees, *Homo erectus*, *Homo floresiensis*, *Homo heidelbergensis*, human beings, Neanderthals
Homo erectus, 92–3, 96, 99, 302; migration, 96
Homo floresiensis, 96
Homo heidelbergensis, 97
Homo sapiens, 100
homosexuality, 64–5
Hooke, Robert, 5–6
hormones, 60
horses, 57, 117, 162
Hoyle, Fred, 329, 391–2 n. 5
Hubble, Edwin, 326
Hulse, Russell, 275–6
human beings, 14, 28, 35, 42, 43, 58, 62, 83, 89, 92, 97, 99, 144, 145, 229, 296; as commodities, 177; DNA, 45, 58, 89–90, 100; diseases, 118; elites, 113, 167; energy use, 162–3; future of, 119; heat shedding, 215; interaction, 95, 98, 110–11, 116, 117, 119, 176–8; interbreeding, 100; newborns, 100; respiration, 196; self-interest, 178; specialisation, 113, 155–6; sweating, 215; *see also* brain, human; hominins
Human Microbiome Project, 16–17
hunter-gatherers, 110, 365 n. 2 *supra*; *see also* gathering
hunting, 93, 95, 101, 107, 109, 116–17; *see also* hunter-gatherers
hurricanes, 216
Hutton, James, 184
Huxley, Aldous, 130; *Brave New World*, 130
Huxley, Thomas, 38
hydrogen, 23–6, 28–31, 207, 208–9, 243, 283–5, 328, 357 nn. 2, 3, 4; 358 nn. 6, 11; 365–6 n. 2, 373–4 n. 2, 377 n. 21, 378 n. 4, 392 n. 6
hydrogen bomb, *see* nuclear bomb
hyperbola, 385 n. 10
hypothalamus, *see* limbic system

Iberian peninsula, 99
IBM, 139, 386 n. 3
ice, 205, 319, 373–4 n. 3
ice, fresh-water, 205
ice ages, 97–8, 99, 107–8, 113, 119, 203–5, 377 n. 22, 387 n. 8; *see also* Snowball Earths
ice caps, 96
ice skating, 317
Iceland, 186
identity, personal, 3, 5
imagination, 79
immigration, 169
immunity, 64, 118
import tariffs, 170–71
incandescence, 345
Incas, 117
incompleteness theorem, 370 n. 4
India, 171
Indonesia, 39
industrial revolution, 114, 115, 176
infinity, 249, 344, 385 n. 10
inflation, 158
information, 7, 40, 71, 76, 81, 82, 84, 144, 221, 281, 297, 308, 309, 350–51, 355 n. 4, 359 n. 8, 380 n. 14, 381 n. 5, 394 n. 14, 395–6 n. 5
information storage, 144
InfoWorld, 148, 371 n. 14
inheritance, 40, 43
Institute for Advanced Study, Princeton, 351
insulation, *see* electricity: insulation
Intel, 147, 148
intelligence, extraterrestrial, 14
interbreeding, 44
interglacial periods, 109
internet, 119
ion channels, 73
ions, 73, 363 n. 6
Iraq, 107
iron, 190, 196, 216, 273, 348
iron oxide, 196
isotopes, 373–4 n. 3
It's Only a Theory, 385 n. 9

Jacob, François, 33
Japan, 387 n. 7
Jarman, Derek, 65
Java, 96
jet stream, 199
jets, cosmic, 347, 349
John Innes Centre and the Institute of
    Food Research, 356 n. 9
Johnson, George, 82
Jones, Steve, 46
Jones, Terry, 368 n. 16
Jupiter, 272, 373 n. 2, 382–3 n. 18; belts,
    198; rotation, 197; zones, 198

Kalahari desert, 93
Kellogg, William, 207
Kelvin, William Thompson, Lord, 218,
    372 n. 4
Kelvin scale, 379 n. 10
Kennedy, Jackie, 49
Kepler, Johann, 271
Keynes, John Maynard, 165
Khufu, Pharaoh, 301
King, Augusta Ada, Countess of Lovelace,
    370 n. 3
Kourou, French Guyana, 197
Krebs cycle, 26, 28
Kuiper Belt, 373 n. 2

labour, 176–7
lactic acid, 25
Laetoli, Tanzania, 91, 120
lakes, 183, 372 n. 2
land, 176–7
Landau, Lev, 339
Lane, Nick, 21
language, 79, 81, 89, 99, 111, 113
Large Hadron Collider, 256, 257, 289, 319,
    347
Larkin, Philip, 63
laser interferometer, 276
lasers, 227, 245, 266–7
'last scattering, epoch of', 302, 388 n. 2,
    389 n. 10, 392 n. 8, 394 n. 14
latitude, 375 n. 11
lava, 186–7, 195

Lavoisier, Antoine, 283
lead, 208, 285
Leakey, Louis, 101
Leakey, Mary, 91
learning, see neurons
Leeson, Nick, 164
Leeuwenhoek, Antonie van, 5, 6
Leibniz, Gottfried, 143
Lemaître, George, 326
leptons, 285–7, 289, 295; indistinguish-
    ability, 287
Lessing, Doris, 82
Lewis, Gilbert Newton, 46
LHB (Late Heavy Bombardment), 195,
    373 n. 2
light, 135, 147–8, 241, 261, 275, 292, 303,
    343, 369 n. 21, 386 n. 14, 388 n. 2; bend-
    ing, 230, 259, 266–8, 272–3; and black
    holes, 384 n. 4; blue, 274; diffraction,
    228; interference, 228, 234; laser, 245,
    276; and matter, 281; oscillation, 274–5;
    as particles, 229–30, 381 n. 4; red, 274;
    reflection, 230–31; speed of, 135, 239,
    247, 249–56, 262, 267, 290, 301–4, 317,
    330, 347, 383 n. 1, 392 n. 9; theory of,
    227–9; transmission, 231; visible, 228,
    327–8, 336, 379 n. 11; as waves, 227–30,
    234, 274, 277, 369 n. 21, 381 n. 4; see also
    energy: light, lasers, light horizon,
    photons, red shift, ultraviolet light,
    wavelength, X-rays
light bulbs, 127, 132, 217, 228, 328, 367 n. 8
light horizon, 337
lightning, 27, 72, 98, 126–8, 130, 367 nn. 6,
    7; 368 n. 18, 368–9 n. 19
LIGO (Laser Interferometer Gravitational
    Observatory), 276
limbic system, 79; amygdala, 79; hippo-
    campus, 79; hypothalamus, 79
lipids (fatty acids), 7
liquids, 296
literacy, 111
lithium, 244, 285
liver, 80
liver cells, 43
Livio, Mario, 377–8 n. 23

llamas, 117

Llinás, Rodolfo R., 77

location, 253, 304, 306, 319, 329, 381 n. 6,
383 n. 3, 395–6 n. 5; dual, 233

logic gate, *see* computers

Lorentz contraction, 251

Lorenz, Edward, 201

Los Angeles, 189

Lovelock, James, 207

LSD (lysergic acid diethylamide), 76

LUCA (last universal common ancestor),
18

lungs, 28–9

Luther, Martin, 53

lysosomes, 11

McGoohan, Patrick, 178

MacLean, Paul, 78

McOwan, Peter, 370 n. 6

McPhee, John, 181

macrostates, 389 n. 9

Madeira, 183

magma, 188

magnesium, 348

magnetic charge: north/south poles, 368
n. 13

magnetic field, 128, 130–31, 133, 134–5, 238,
244, 369 nn. 20, 21; curl, 130

magnetic force, 291

magnetism, 134–5, 186, 369 n. 21

magnets, 129–31, 134, 238

Maguire, Eleanor, 84

malaria, 43, 55, 64

Malaysia, 39

Maldacena, Juan, 351

mammals, 112

Manhattan Project, 175

maps, 185

Margulis, Lynn, 10

market, the, 167–78; complexity, 175–6;
deregulation, 170, 371 n. 2; equilibrium,
172–4; fluctuations, 173–4; instability,
172–3, 176; regulation, 168–9; *see also*
foreign exchange, futures market,
import tariffs, sanctions, shares, stock
market

marketplace, 155, 157

Mars, 377–8 n. 23, 382–3 n. 18

Marsden, Ernest, 284, 286

Marx, Karl, 167

mass, 255–8, 262–4, 266, 268, 270–73, 283,
288–91, 313–14, 317, 328, 334–6, 345–6,
383–4 n. 5, 385 n. 9, 394 nn. 15, 16

mass production, 114

mathematics and mathematicians, 142–3,
173–4, 200–201, 270–71, 297, 305,
315–16, 318, 326, 370 n. 4, 379 n. 6,
390 n. 5

Matrix, The, 369 n. 2

matter, 222, 230, 270–71, 275, 281–2, 286,
290, 295, 329, 332, 335–7, 343, 347, 380
n. 16, 385 n. 8, 387 n. 12, 389 n. 10, 392
n. 8, 393 n. 11, 394 nn. 14, 16; cosmic,
274; theory of, 227–8; *see also* dark
matter

Maxwell, James Clerk, 134–5, 282, 291, 369
n. 20

Maxwell's theory, *see* electromagnetism,
Maxwell's theory of

MC1R, 59

Mears, Ray, 229

measles, 117, 118

meat, 93, 94, 99, 112–13

mechanisation, 114

Médaille, John, 161

meditation, *see* neurons

meiosis, 57, 61, 62, 362 n. 15

memory, 111; *see also* computers, neurons

Mendel, Gregor, 40–42

meninges, 79

meningitis, 79

menopause, 65–6

Mercury, 209, 271–3; orbit precession,
272–3

mesons, 291

meteorites, 184

methane, 205

Mexico, Gulf of, 201, 204

Michelson, Albert, 249

microchips, 144, 146–8

microorganism, 9, 14, 17, 357 n. 15

microstates, 389 n. 9

microwaves, 327–8, 391 n. 3
mid-Atlantic Ridge, 186, 187
Middle East, 96
Milanković, Milutin, 97
Milanković cycles, 97, 203, 364 n. 8
military, 110
Milky Way Galaxy, 5, 14, 57, 315, 325–7, 334–5, 343, 345–6, 348, 349, 359 n. 4, 394 n. 15
millet, 112
Milne, A. A., 299
mimiviruses, 45
Minkowski, Herman, 254, 305
Minsky, Marvin, 83
mitochondria, 10–11, 361–2 n. 13
mitosis, 57, 58
mobiles phones, 148
molecules, 297; and cell-building, 7–8, 27; copying, 51–2; electrically charged, 137, 363 n. 6; formation, 24, 145, 243, 285, 295–6, 378 n. 4; and heat, 216–18, 378–9 n. 5, 380 n. 14; high-altitude, 204; splitting, 204, 205, 207
momentum, 293, 382 nn. 14, 15; 384 n. 5; law of conservation of, 317; law of conservation of angular, 317
money, 151–64, 177; divisibility, 157–8; IOUs, 159; portability, 158
monogamy, 95
Monty Python, 368 n. 16
Moon, 89, 97, 268, 301, 330, 384–5 n. 7, 395–6 n. 5; crust, 373 n. 2; gravity, 263, 269, 314; impact basins, 372–3 n. 5, 373 n. 2; orbit, 314; Sea of Crises, 120
Moore, Gordon, 148
Morgan, Thomas Hunt, 41
Morley, Edward, 249
morphogens, 359 n. 7
Morse code, 239
mortgages, 160
mosquito, 124–5, 366 n. 4
motion, laws of, 251, 314
Mount St Helens, 187
mountains, 188
mouth-to-mouth resuscitation, 28
MRSA (Staphylococcus aureus), 16

MS (multiple sclerosis), 363–4 n. 17
mud, 183–4
mudstone, 184
multicellular organisms, 13, 63, 72, 74
multiverse, 359 n. 4, 390 n. 5
muons, 252–3, 287–8, 383 n. 2; see also neutrinos
muscles, 5, 74, 76, 81, 83, 84, 216, 217, 255; power, 114
mushrooms, 45
myelin, 363–4 n. 17

n-type material, 146
NASA (National Aeronautics and Space Administration), 197, 394 n. 18
National Grid, 133
National Institute of Standards Technology, 266
natural selection, 37–40, 44, 46–7, 52, 55, 62, 63, 85, 91, 108, 111, 155, 176, 189, 358–9 n. 2; see also inheritance, traits, variation
Navier–Stokes equation, 200
Neanderthals, 97, 99–100; interbreeding, 100
nematode worm, 78
neocortex, 79; corpus calossum, 79; hemispheres, 79; left side/right side, 79–80
Neptune, 373 n. 2
nerve cells, see neurons
nervous system, 72, 77, 78, 363 n. 13
Netherlands, 171
neurons, 5, 15, 43, 74–8, 80–85, 355 n. 1, 363 n. 13, 363–4, n. 17; axons, 74–6, 80, 82, 363 n. 8, 363–4 n. 17; connections, 80–83; dendrites, 75–7, 80, 82, 83; forgetting, 83; learning, 80–85; meditation, 84; memory, 76, 80–84; networks, 76, 83–4; receptors, 75, 77; repetition, 81; synapses, 75, 77; synaptic gap, 82, 363 n. 8; terminal buttons, 363 n. 8; see also neurotransmitters
neuroplasticity, 83, 85
neuroscience and neuroscientists, 78, 80, 86, 296
neurotransmitters, 60, 75–7, 82; see also

acetylcholine, dopamine, glutamate, serotonin

neutrinos, 231, 287–8, 386 n. 14, 387 nn. 7, 8, 12; electronneutrinos, 285–7; muon-neutrinos, 287; sterile, 387 n. 12; tau-neutrinos, 288

neutron stars, 115, 275–7, 386 n. 12, 395 n. 3

neutrons, 231, 256, 283, 285–6, 288–91, 358 n. 6

New Scientist, 274

New York, 132

New York Times, 318

Newton, Isaac, 115, 143, 231, 261–2, 270–73, 286, 313–15, 385 nn. 8, 9; 386 n. 2, 390 n. 3, 395–6 n. 5; *see also* gravity

nitrogen, 8, 195

Nobel Prize, 135, 173, 316; for Physics, 235–6, 276, 281, 344, 370 n. 9, 386 n. 3

Noether, Emmy, 316, 318, 379 n. 6

non-locality, 239, 243; *see also* location

North America, 96, 97, 117; *see also* Americas

North Atlantic conveyor, 205

North Pole, 96

nothingness, 319

nuclear bomb, 125, 257–8, 327, 331

nuclear forces, 289; strong, 256, 285, 286, 289–90, 392 n. 6; weak, 289, 291

nuclear fuel, 273, 387 n. 7

nuclear power, 131, 163, 368 n. 16

nuclear radiation, 43

nuclear reaction, 377 n. 21

nuclear reactors, 115, 227

nylon, 127, 367 n. 9

oceanic crust, 188–90

oceans, *see* seas

oil, 31, 131, 172, 206

olive, 109

Ophioglossum vulgatum, 57

optics, 251

orang-utans, 92

orbitals, 242–4

organelles, 11–13, 61

Ørsted, Hans Christian, 129

outbreeding, 62

ovaries, 60, 66

ovum/ova, 15, 58, 63, 66, 361–2 n. 13

oxidation, 357 n. 4

oxygen, 23–31, 119, 195–6, 205, 207, 233, 357 nn. 2, 3, 4; 378 n. 4

ozone, 204

p-type material, 146

Pacific Ocean, 96, 97, 204

palaeoanthropology, 91

panspermia, 360 n. 2

parabola, 385 n. 10

parasites, 47, 55–6, 99, 460 n. 9

Paramecium, 74

Paris, 239

particle accelerators, 292–3

particle physics, 286, 331, 336

particles, subatomic, 245, 252, 257, 285, 287, 293, 319, 335, 350; *see also* antiparticles, leptons, muons, quarks, taus

Pauli Exclusion Principle, 244–5, 287, 383 n. 19, 387 n. 11, 387–8 n. 15

pea plants, 40–42, 109

peacock, 358 n. 1

pelvis, 91

Penrose, Roger, 305, 389 n. 10

personality, 79

pharmaceutical industry, 176

philosophy, 114, 115

phosphorus, 146

photino, 292

photography, 367 n. 6

photolithography, 147

photons, 21, 30, 126, 222, 230–31, 237, 241, 242, 245, 267, 289, 292, 294, 328, 380 n. 12, 384 n. 5, 389 n. 10, 392 n. 8, 394 n. 14; virtual, 368 n. 11

photoresistance, 147, 371 n. 11

photosynthesis, 29–31, 195, 205, 356 n. 6; artificial, 31

physics and physicists, 18, 24, 125, 128–30, 134, 135, 144, 172, 174, 221–2, 227, 229, 231, 235, 238, 239, 249, 251, 257–8, 259, 261–2, 266, 273, 274, 276, 279, 281, 283, 286, 289–91, 294–7, 306–9,

311–19, 330, 331, 333–4, 339, 344, 350, 351, 370 n. 5, 375 n. 10, 379 n. 10, 383 n. 2, 383–4 n. 5, 386 n. 2, 392 n. 7, 395–6 n. 5; *see also* particle physics

physics, laws of, 217, 219, 255, 286, 306, 307, 332, 344, 379 n. 6, 388 n. 8, 389 n. 9

Picasso, Pablo, 141

pigs, 33, 57, 112, 118

Pink Floyd, 3, 341

Pinker, Steven, 111

Pisa, Tower of, 263

piston, 215–19, 282, 378–9 n. 5, 379 n. 8, 380 n. 14

pit viper, 47

Planck's constant, 382 n. 9

planetesimals, 372–3 n. 5

planets, 289; formation, 372–3 n. 5; orbits of, 23, 382–3, n. 18; shape, 365–6 n. 2

plants, 6, 14, 18, 25, 29–31, 77, 93, 116, 221, 296, 365 n. 2 *supra*; poisons, 99; reproduction, 37, 38, 53; *see also* farming

plate tectonics, 185–91, 373 n. 7

plutonium, 168, 283

Polanyi, Karl, 176–7; *The Great Transformation*, 176

Polaris, 376 n. 17

politics, 168, 169, 172, 177

pollen, 356 n. 11

pollution, 171

population growth, 110–12, 116, 117

potassium, 189, 244

poverty, 162–3

power stations, 131–3, 163, 368 n. 17; *see also* nuclear power

precession, 376 n. 17

predictability, 201

pre-evolution, 52

pregnancy, 67

pressure, 228, 282, 297, 393 n. 11; *see also* air: pressure

primates, 79, 95

Principe, 268

Prisoner, The, 178

privatisation, 371 n. 2

Proceedings of the Natural History in Brünn, 41

profit, *see* capitalism: profit

prokaryotes, 6, 8, 9, 10–12, 16, 17, 361 n. 10; *see also Thiomargarita namibiensis*

protectionism, 170–71

proteins, 6, 15, 17, 28–30, 42, 58, 59, 72–3, 89, 99, 358 n. 5; 361 n. 10; broken, 59; complexes, 26; expression, 8–9, 13, 42, 46, 52, 60; machinery, 58; modification, 11; and RNA, 355 n. 4; shape change, 358 n. 7; structure, 6–7; transporter proteins, 7; *see also* histones, ion channels, MC1R, tubulin

protons, 27, 124, 256, 283, 285–6, 288–91, 358 nn. 6, 7; 365–6 n. 2, 366 n. 3, 387 n. 7; positive electric charge, 283, 285

protozoans, 16

Prout, William, 283

PSR B1913+16 (binary pulsar), 276

public services, 371 n. 2

Publius Scipio, 301

Pugh, Emerson M., 85–6

pulley, 220–21

pulsars, 275–6, 395 n. 3; *see also* PSR B1913+16

PVC, 368 n. 17

Pyramids, 301, 376 n. 17

quantum electrodynamics, 281

quantum fluctuations, 335

quantum numbers, 242, 244

quantum spin, 237–8, 244; *see also* angular momentum

quantum theory, 225–45, 289, 293–4, 318, 332, 334, 337, 339, 366 n. 3, 368 n. 11, 393 nn. 10, 12; 393–4 n. 13, 395–6 n. 5; computation, 236; unpredictability, 231–2, 238, 381 n. 5; probability, 232, 233

quantum waves, 236, 240, 242, 293, 381 n. 6, 381–2 n. 8; *see also* orbitals, waves: interference, waves: superposition

quantum weirdness, 234, 236–7, 240

quarks, 245, 256, 285–7, 289–91, 295, 296, 297, 382 n. 10; bottom-quarks, 288; charm-quarks, 287; down-quarks, 286–7, 289; indistinguishability, 287;

strange-quarks, 287; top-quarks, 288; up-quarks, 286–7, 289; see also anti-quarks
quasars, 345–6, 348, 349
Queen Mary, University of London, 370 n. 6

Rabi, I. I., 288
radio communication, 135
radio waves, 135, 275, 351, 385 n. 11
radioactive dating, 184
radioactivity, 189, 381 n. 5
radium, 384
Raglan, Fitzroy Somerset, 4th Baron, 105
rain, 163, 193, 207
Ramón y Cajal, Santiago, 86
Raskin, Jef, 369 n. 1
Raspall, François-Vincent, 17
receptors, see neurons
red dwarfs, 348
red giants, 209, 377–8 n. 23
Red Queen Hypothesis, 56
red shift: cosmological, 274; gravitational, 274–5
reduction, 357 n. 4
relativistic aberration (or beaming), 383 n. 1
relativity, 304–5, 308; covariance, 391 n. 9; general theory of, 259–77, 303, 317, 343, 344, 384 n. 6 infra, 390 n. 3, 392 n. 9, 393–4 n. 13; special theory of, 239, 247–58, 262, 265, 267–8, 303, 317, 392 n. 9
Resnikoff, Howard, 221
respiration, 21–31
ribosomes, 8, 11
rice, 112
RNA (ribonucleic acid), 8, 11; information storage, 355 n. 4; and proteins, 11, 355 n. 4; replication, 355 n. 4
rocket, 261, 263–7, 270; propulsion, 23–4, 26, 31, 357 n. 2
rocks, 181, 183–4, 186, 187, 215; iron-rich, 196; magnetism, 186; radioactive, 189; sedimentary, 184, 372 n. 2; volcanic, 187–8
Rohrer, Heinrich, 386 n. 3

Royal Belgian Institute of Natural Sciences, 365 n. 6
Royal Society, London, 5
Rumsfeld, Donald, 336
Russia, 360 n. 1
Rutherford, Adam, 15
Rutherford, Ernest, 284
Ryan, Robert T., 200

Sackmann, Juliana, 377–8 n. 23
Sagan, Carl, 10, 195, 207, 323
Sagittarius A*, 348
salt, 6, 73, 201, 204; as currency, 158
San Andreas Fault, California, 174, 189
San Francisco, 189
sanctions, 170
SatNavs, 147
Saturn, 373 n. 2; rings, 144
Scarre, Chris, 107, 111
scavenging, 93, 95
science, 114–15, 129
Schmidt, Maarten, 345
Schulman, Larry, 388 n. 8, 389 n. 10
Scotland, 184, 235
Scott, Dave, 263
sea ice, 201, 204
sea squirt, 78
sea water, 96; see also sea ice
seabed, 185–8, 190
seas, 190, 201–3, 207, 208, 372 n. 2, 373–4 n. 3; heat, 201; salt, 201
seashells, 183, 188, 190
Seinfeld, 79
seismic tomography, 190
selectron, 292
semiconductors, 146
senses, 71, 76, 84
sequoias, 35
serotonin, 76
Serratia marcescens, 56
settlements, 109–11, 119; defence, 110
sex cells, see gametes
sexual differences, 58, 60
sexual reproduction, 43–4, 49–67, 360 n. 9
sexuality, 65
Shakespeare, William, 130, 213, 295

shares, 169
sheep, 37, 112
Shipman, Pat, 113
ships, 114, 115
Shockley, William, 370 n. 9
sickle cell anaemia, 64–5
silicon, 83, 145, 146–7, 363 n. 13
silkworms, 112
silver, 368 n. 15
Sinclair, Upton, 167
singularity, *see* black holes
skull, 79, 94, 96
slime moulds, 362 n. 19
smallpox, 117, 118
Smith, Adam, 168, 172, 178; *The Wealth of Nations*, 168
snooker, 317
Snow, C. P., 213
Snowball Earths, 205
sociology and sociologists, 148, 296
sodium, 73, 244, 273
Sokal, Alan, 316
solar flares, 174
Solar System, 209, 241, 325, 382–3 n. 18; formation, 184, 372–3 n. 5, 373 n. 2; gravity, 270–71; outer, 373 n. 2
soldiers, *see* military
solids, 296
Sommerfeld, Arnold, 222
sonar surveying, 186
sound waves, 277
South Africa, 98
South America, 36–7, 96, 97, 117, 185–8, 204, 205; *see also* Americas
South Sea Bubble, 371 n. 2
space, 134, 157–8, 233, 242, 250, 253–4, 256–8, 261, 297, 301, 302, 305, 317, 318, 329, 333, 337–8, 341, 344, 381 n. 4, 383–4 n. 5, 388 n. 3, 391 n. 4, 392 nn. 8, 9; curvature, 268, 270; shells, 302, 303; *see also* location: dual, space–time
Space Telescope Science Institute, Baltimore, 377–8 n. 23
space–time, 253–4, 256, 258, 269, 275, 277, 305, 306, 308, 317, 329, 343, 381 n. 4, 384 n. 6 *supra*, 384 n. 4; warpage

(geometry), 269–70, 273, 275, 385 n. 8
species, 44–5
speed, 217, 250–51, 378–9 n. 5; infinite, 249, 255; uniform, 261, 265; warped, 268; *see also* light: speed of
sperm, 15, 58, 63, 361 n. 6
spinal cord, 5
spindle, 130
sponges, 72, 363 n. 5
Springsteen, Bruce, 93
SRY (Sex-Determining Region of the Y chromosome), 59–60
stag, 358 n. 1
stagnation, 170
standard of living, 162–3, 171
Standard Model, 290
Stanford University, California, 351
Staphylococcus aureus, *see* MRSA
Star Trek, 71, 128, 157
starfish, 316, 391 n. 9
starlight, 222, 328
Starobinsky, Alexei, 331
stars, 5, 16, 18, 57, 78, 220, 222, 262, 273–4, 277, 289, 292, 314, 325–6, 328, 335–7, 345–51, 374 n. 7, 380 n. 13, 384 n. 4, 386 n. 12, 389 n. 10; *see also* black holes, neutron stars, supernovae, white dwarfs
steady state, 391–2 n. 5
steam, 215–18, 227, 282, 378 n. 4, 378–9 n. 5, 379 n. 7. 380 n. 14; power, 131
steam engine, 213, 215–16, 218, 219, 282, 347 n. 1, 379 n. 7
steel, 275
stellar winds, 277–8 n. 23
Stenger, Victor, 319
Stephens, Mark, *see* Cringely, Robert X.
Stiglitz, Joseph, 169; *The Price of Inequality*, 169
STM (Scanning Tunnelling Microscope), 386 n. 3
stock market, 168, 173, 174; *see also* shares
stomach, 16, 99; lining, 15
Stoppard, Tom, 284
stratosphere, *see* Earth: stratosphere
string theory, 293–4
strokes, 80, 85; rehabilitation, 85

stromatolites, 8

subsidy, 170–71

sugars, 26, 30; *see also* glucose

sulphuric acid, 208

Sumerians, 111

Sun: eclipse, 268, 272, 384–5 n. 7; envelope, 377–8 n. 23; flares, 376–7 n. 18, 377 n. 19; formation, 372–3 n. 5; future, 377–8 n. 23; heat output, 203–4, 208–9, 377 n. 21, 382 n. 15; luminosity, 208–9, 377 nn. 21, 22; magnetic field, 376–7 n. 18; mass, 268, 272, 347, 348; as nuclear reactor, 200, 286; neutrinos, 286; orbit, 325; rotation, 376–7 n. 18; solar cycle, 376–7 n. 19; sunspots, 376–7 n. 18; temperature, 191, 220, 377 n. 19; tidal bulge, 377–8 n. 23; ultraviolet radiation, 204

Sunday Times, The, 164

sunlight, 30–31, 163, 203, 204–6, 221, 285, 377 n. 21, 378 n. 1, 387 n. 8, 392 n. 6

supercomputers, 78

SuperKamiokande detector, 387 n. 7

supernovae, 252, 277, 386 nn. 12, 14

supersymmetry, 291–2

supply and demand, 155, 158, 164, 167, 170, 171

survival, 37, 38, 46, 53–4, 57, 62, 65, 72, 73, 89, 91, 100–101, 309, 358 n. 1

Susskind, Leonard, 351

sweating, 92, 215

symbiosis, 296

symmetry, 311, 316–19, 379 n. 6; rotational, 316, 317; timetranslation, 317, 379 n. 6; translational, 317

synapses, *see* neurons

synthesis, 296

Syria, 107

Sze, Arthur, 297

taus, 288; *see also* neutrinos

taxation, 371 n. 2

taxi drivers, 84, 169

Taylor, Joseph, 275–6

tectonic plates, 187–91; *see also* plate tectonics

teeth, 94, 99

Tegmark, Max, 390 n. 5

telecommunications, 119, 135

telescopes, 301, 302, 336, 338

television, 135, 327

temperature, 217–19, 282, 291, 297, 319, 328, 330, 378 n. 3, 379 nn. 7, 8, 10; 380 n. 13, 392 n. 7; *see also* absolute zero, body temperature, Celsius scale, Earth: temperature, Kelvin scale, Sun: temperature

Terry, John, 95

Tesla, Nikola, 132

testes, 60

testosterone, 60

thermodynamics, 213–23, 306; first law of, 217; second law of, 213, 217–21, 307, 388 n. 8, 389 n. 9

Thiomargarita namibiensis, 9

Thomas, Lewis, 3, 5, 10, 18, 21, 45, 52

Thomson, George, 235–6, 381–2 n. 8

Thomson, J. J., 236

thorium, 189

thunderstorm, 27, 125, 128, 367 n. 9

thymine, 42

Tigris river, 107

time, 134, 157–8, 250–54, 256–8, 261, 265, 266, 273, 299–309, 317, 318, 329, 337, 341, 343–4, 345, 381 n. 4, 383 n. 3; 'arrow', 307–8; dilation, 251, 266; *see also* space–time

tools, 90, 91, 93–6, 98, 99, 116

trade, 114, 153–8, 162–4, 167, 169; international, 161–2; and space, 157–8; and time, 157–8; *see also* exchange rate, marketplace, supply and demand

trade winds, 198

trading circle, 154

traits, 40, 42–4, 54

transatlantic cables, 185–6, 372 n. 4

transformers, 133

transistors, 77, 144–9, 370 nn. 6, 9

trees, 30, 44, 51, 53, 91, 92, 107, 163, 378 n. 1

troposphere, *see* Earth: troposphere

Tsiolkovsky, Konstantin, 209

tuberculosis, 118

tubulin, 12
Turing, Alan, 139, 142–3, 370 n. 4
Turkey, 107
Tutu, Desmond, 95, 364 n. 6
Twain, Mark, 118, 203
Twitter, 381 n. 7, 382 n. 17
Tyndall, John, 196
Tyson, Neil deGrasse, 316

Uhuru satellite, 345
ultraviolet light, 43, 148, 207, 377 n. 19
uncomputability, 370 n. 4
undecidability, 370 n. 4
United States, 188
Universal Turing Machine, 142–3
Universe: age, 149, 308, 337, 345, 394 n. 18;
    and dark energy, 292, 333–6; and
    entropy, 219–22, 308, 380 n. 12; expan-
    sion, 222, 274, 326–31, 333–4, 338, 380
    n. 16, 392 n. 9, 394 n. 18; formation, 288,
    290, 295, 308, 319, 325–32, 335, 393–4
    n. 13, 394 nn. 14, 16; future of, 266, 306,
    344; and galaxies, 78, 359 n. 4; as image,
    78, 86; inflation, 331, 333, 337–8, 381
    n. 5; and light, 222, 249–50, 267, 277,
    302, 328, 330, 345, 386 n. 14, 388 n. 2;
    and mass energy, 292, 333; observable,
    18, 302, 337, 392 n. 9, 394–5 n. 19; past
    of, 274, 301–3, 326, 346, 389 n. 10, 391–2
    n. 5, 392 n. 8; perception of, 71, 86, 144,
    218, 297–8, 309, 315, 325, 330, 333, 336,
    343, 349, 352, 370 n. 5, 390 n. 4, 391 n. 4;
    and quantum theory, 227, 232, 239, 243,
    257, 275, 294, 315, 335, 337, 339, 393
    n. 12; scale, 326, 392 n. 9; temperature,
    220–21, 331–5, 380 n. 13; see also big
    bang, cosmic background radiation,
    event horizon, galaxies, Hawking
    radiation, 'last scattering', multiverse,
    red shift
University College, London, 84
uranium, 189, 243, 283
Uranus, 373 n. 2, 374 n. 7
US Navy, 186
uterus, 63

vacuum, 237, 257, 301, 331–4, 350, 393
    n. 10; false, 332, 393 n. 11; inflationary,
    332, 334, 335, 395–6 n. 19
Valéry, Paul, 319
Van Valen, Lee, 55–6
variation, 40, 43, 45, 62
Velcro, 309
velocity, 317, 374–5 n. 9
Venus, 190, 196, 207–8, 209, 325, 373 n. 7;
    day/year, 374 n. 5; orbit, 374 n. 5; water,
    207
Very Large Array, New Mexico, 347
vesicles, 8
Victoria, Queen, 372 n. 4
Vienna, 389 n. 9
viruses, 18–19, 43
void, 318, 319
volcanoes, 186, 188, 190, 191, 195, 207, 208;
    eruptions, 184, 187, 205; see also lava,
    magma
voltage, see electric current: voltage
voltage-gated ion channels, see cells
volume, 228, 382 n. 15
vortices, 199

wages, 164, 169
Wallace, Alfred Russel, 39, 358–9 n. 2
war, 111, 113
water, 26, 29, 30–31, 183, 184, 187–9, 193,
    195, 204, 205, 208, 296, 319, 359–60 n. 9;
    363 n. 5, 368 n. 16, 373 n. 7; boiling, 189,
    201; condensation, 215; erosion, 184;
    molecular structure, 24; power, 114, 131,
    163; swirl direction, 199; ultrapure, 387
    n. 7; vapour, 28, 193, 195, 206–8, 391
    n. 4; see also evaporation, ice: fresh-
    water, ice, sea ice, steam, wetness
water wheel, 132
Watson, James, 85, 359 n. 6
Watson, Thomas J., 139
wavelength, 148, 240–41, 382 n. 14
waves, 240; gravitational, 275–7; inter-
    ference, 234–7, 381–2 n. 8; super-
    position, 233, 236–8, 243, 381–2 n. 8;
    see also light: as waves, quantum waves,
    radio waves, sound waves

weapons, 91, 93, 117
Weather Girls, 193
weather forecasting, 200–201
weather systems, 199–200; anticyclones, 199–200; cyclones, 199
weathering, 184
Wegener, Alfred, 185, 187, 189
Weyl, Herman, 316
Weinberg, Steven, 316, 344
wetness, 296
whales, 35; killer whales, 65; short-finned pilot whales, 66
wheat, 109
Wheeler, John, 144, 270, 309, 318, 344, 395 n. 3
white dwarfs, 273–5
Wigner, Eugene, 315
Wilde, Oscar, 71, 153, 177
Wilkinson Microwave Anisotropy Probe, 394 n. 18
William the Conqueror, 301
Williams, Tennessee, 111
Wilson, Edward O., 78, 101
wind: erosion, 184; high-altitude, 198–9; power, 131, 163; see also jet stream, stellar winds, trade winds

windmills, 163
Winnipeg, Canada, 205
wire coil, conducting, 129–31, 133, 135, 368 n. 15
Woese, Carl, 355–6 n. 5
wolf, 93
Wolpert, Lewis, 359 n. 7
womb, see uterus
work, 216–21, 357 n. 1, 379 n. 7
World War I, 273
World War II, 142
World Wide Web, 80, 164
worms, intestinal, 55
Wright, Stephen, 344
writing, 111, 113
WTO (World Trade Organisation), 170

X-rays, 148, 190, 345
xenophobia, 111

yeast, 25
Young, Patrick, 201
Young, Thomas, 234

Zurich, Switzerland, 386 n. 3
zygotes, 58, 362 n. 15

Also by Marcus Chown

ff

Quantum Theory Cannot Hurt You: Understanding the Mind-Blowing Building Blocks of the Universe

The entire human race would fit in the volume of a sugar cube. You age faster at the top of a building than at the bottom. Every breath you take contains an atom breathed out by Marilyn Monroe. All of these are true – but why? Two brilliant ideas – quantum theory and Einstein's general theory of relativity – hold the key. Marcus Chown gives us a fascinating, accessible and witty introduction to these two theories that underpin all of modern physics. It's a book you can read in a morning, but which will leave you excited about science for years to come.

'Chown discusses special and general relativity, probability waves, quantum entanglement, gravity and the Big Bang, with humour and beautiful clarity.' *Guardian*

'Readers will experience happy eureka moments.' *The Times*

'A must-read for anyone who wants to better understand this crazy universe we live in. Superb.' *Astronomy Now*

ff

We Need to Talk About Kelvin: What Everyday Things Tell Us About the Universe

Acclaimed popular science writer Marcus Chown shows how our everyday world reveals profound truths about the ultimate nature of reality. The reflection of your face in a window tells you that the universe at its deepest level is orchestrated by chance. The static on a badly turned TV screen tells you that the universe began in a big bang. And your very existence tells you this may not be the only universe but merely one among an infinity of others, stacked like the pages of a never-ending book . . .

'Perfect for someone who wants a non-intimidating intro to modern physics, or a precocious teenager who won't stop asking why.' *New Scientist*

'Chown writes with ease about some of the most brain-bending of concepts and makes you really think about science.' BBC *Focus Magazine*

'Chown writes very fluently, helping us to visualise things with matchboxes and Lego bricks.' *Guardian*

ff

The Never-Ending Days of Being Dead: Dispatches from the Front Line of Science

Did you ever wonder . . . where did we come from, and what the hell are we doing here? Is Elvis alive and kicking in another space domain? What's beyond the edge of the Universe? Did aliens build the stars? Can we live forever? Acclaimed popular science writer Marcus Chown takes us to the frontier of science, revealing that the questions asked by today's most daring and imaginative scientists are in fact those very ones which keep us up at night. An ambitious yet superbly readable exploration of the mysteries of the Universe.

'We must all be grateful to writers like Chown who are able to make accessible work that in its crude form is not only inaccessible to outsiders, but unknown to them.' *Independent on Sunday*

'I can not find fault with this book, the style is yummy, the mathematics non-existent and the concepts surprising.' *Astronomy Now*

'A limousine among popular-science vehicles.' *Guardian*

ff

Afterglow of Creation: Decoding the Message from the Beginning of Time

It's in the air around you. It carries with it a baby photo of the Universe. Its discoverers mistook it for pigeon droppings yet still won the Nobel Prize. *Afterglow of Creation* tells the story of the biggest cosmological discovery of the last hundred years: the afterglow of the big bang. The result of this find was a 'baby photo' of the universe – sensationally described as 'like seeing the face of God' – which revolutionised our picture of the cosmos. Marcus Chown goes behind the initial hysteria to provide a lively and accessible explanation of this hugely important research – and gives us a compelling and exuberant tale of the human side of science.

'Superbly captures the spirit of scientific discovery.' *Sunday Times*

'A very good piece of storytelling.' *New Scientist*

'A "science for dummies" take on creation, built around an account of the discovery of radiation ripples from the Big Bang.' *The Times*